LATTICE HIGGS WORKSHOP

LATTICE HIGGS
WORKSHOP

FLORIDA STATE UNIVERSITY
MAY 16 — 18 1988

Editors

B BERG
G BHANOT
M BURBANK
M CREUTZ
J OWENS

World Scientific
Singapore • New Jersey • Hong Kong

Published by

World Scientific Publishing Co. Pte. Ltd.
P. O. Box 128, Farrer Road, Singapore 9128

U. S. A. office: World Scientific Publishing Co., Inc.
687 Hartwell Street, Teaneck NJ 07666, USA

LATTICE HIGGS WORKSHOP

ISBN 9971-50-686-6

Printed in Singapore by JBW Printers & Binders Pte. Ltd.

PREFACE

First, one should look back and remember...

During the period May 16–18, 1988, over fifty scientists from around the world participated in the *Lattice Higgs Workshop* held at The Florida State University, Tallahassee, Florida. Many people contributed to the success of this workshop. Professor Joseph Lannutti, the Director of the Supercomputer Computations Research Institute (SCRI), made everything possible with his support. Pat Meredith did most of the hard organizational work, from getting the poster prepared to mailing off the per diems. Without her sustained efforts, things would have broken down in many embarassing ways. Susan Lampman and the staff of the Florida State University's Center for Professional Development must be highly commended for a flawless performance. They created an impression that all the arrangements at the Conference Center were effortless—a testimony to their professional competence. The SCRI staff helped us also in many subtle ways that were essential to the smooth running of the workshop. We acknowledge financial assistance for the workshop from SCRI and the Department of Energy.

The work of 'repairing' the manuscripts into publishable form was done by Donna Middleton and Jeanne Waggaman. They performed this difficult task with an enthusiasm and a persistance that made it seem easy.

Finally, the speakers and participants deserve thanks: the speakers for their excellent talks and their understanding when pushed for their contributions to these proceedings, and the participants for their constant interest through a long schedule of talks. Without them, in a very literal sense, the workshop would not have been possible. M. Tsypin from the Lebedev Institute was unable to attend but asked us to include his contribution in these proceedings.

If our memory has been faulty, and in remembering, we have forgotten someone who helped, we would like to thank them the most. Their contribution must indeed have been crucial, like the air one breathes but is unaware of.

The papers collected in this volume represent the current state of the subject of lattice methods applied to Higgs phenomenology. As was clear from the conference, there are still many open issues in this field which are of relevance to supercollider physics. We hope that the present conference is only the first of a series which will bring together high energy lattice enthusiasts, experimentalists and phenomenologists.

July, 1988   The Editors

TABLE OF CONTENTS

viii

LATTICE HIGGS WORKSHOP – CONFERENCE SUMMARY

Stephen L. Adler

The Institute for Advanced Study
Princeton, New Jersey 08540

We have just heard a program of very good talks, and what with the beaches being far away, everyone has heard all the talks! Hence, I will not attempt a talk-by-talk synopsis, but rather will make some brief historical and summary remarks and then, looking towards the future, will discuss the question of improved computational algorithms for some of the models that have been considered here.

Turning first to the keynote theme of the conference, "bounding the Higgs," we recall some formulas from the seminal paper of Dashen and Neuberger.[1] Let $v = \langle \phi \rangle$ be the vacuum expectation value of the Higgs scalar, λ_R the renormalized ϕ^4 coupling, and g_R the renormalized gauge coupling; in terms of these, the Higgs mass m_H and the charged intermediate boson mass m_W are given by

$$m_H^2 = 2v^2 \lambda_R, \quad m_W^2 = \frac{1}{4}v^2 g_R^2. \tag{1}$$

Solving for v in terms of the Fermi constant G_F we have

$$v = \left(\frac{g_R^2}{4m_W^2} \right)^{-1/2} = \left(\sqrt{2} G_F \right)^{-1/2} = 246 GeV. \tag{2}$$

The conjectured triviality of ϕ^4 field theory suggests that λ_R can be estimated by the one-loop-renormalization group, giving

$$\lambda_R \sim \frac{1}{\frac{3}{2\pi^2} \ln(\Lambda/\mu)}, \tag{3}$$

with Λ an ultraviolet cutoff and μ an effective subtraction point. (Although μ is in principle fixed in terms of v, the precise proportionality constant relating μ and

v can only be determined by a Monte Carlo simulation.). Combining Eqs. (1)–(3) gives the estimate

$$m_H \sim \frac{2\pi}{\sqrt{3}} \frac{246 GeV}{(\ln \Lambda/\mu(^{1/2}} \sim \frac{900 GeV}{(\ln \Lambda/\mu)^{1/2}}. \tag{4}$$

If m_H is small compared to $900 GeV$, the energy scale Λ associated with the breakdown of the ϕ^4 effective theory and the appearance of new physics is very large. However, as Λ decreases m_H increases; at the "crossover" where $m_H \sim \Lambda$, the Higgs mass itself lies in the "new physics" regime and the model has broken down. Thus, the crossover point determines an effective upper bound for the Higgs mass, in the sense that if the Higgs is not found below the upper bound, new physics is sure to exist not far above the upper bound.

Historically, the first attempt at a Monte Carlo evaluation of the upper bound was made by my graduate student Charles Whitmer in 1984.[2] (A very similar calculation was carried out independently by M.M. Tsypin in 1985).[3] Whitmer's method consisted of using the effective potential $U_\Omega(\bar{\varphi})$ which governs the distribution of

$$\bar{\varphi} = \frac{1}{\Omega} \left| \sum_x \phi(x) \right|, \quad \Omega = \text{lattice volume}, \tag{5}$$

through the formula

$$P(\bar{\varphi}) = N \exp\left[-\Omega U_\Omega(\bar{\varphi})\right], \tag{6}$$

provided there are no significant wave function renormalization corrections, v and m_H are related to the effective potential by

$$v = \langle \bar{\varphi} \rangle = \int d\bar{\varphi}\, \bar{\varphi}\, P(\bar{\varphi}),$$
$$m_H^2 = \left.\frac{\partial^2}{\partial \bar{\varphi}^2} U_\Omega(\bar{\varphi})\right|_{\bar{\varphi}=v}. \tag{7}$$

A calculation on up to 9^4 lattices used the two point functions to estimate the wave function renormalization and gave

$$Z \approx 1 \tag{8}$$

to within five percent; defining the "crossover" as the point where the inverse lattice spacing a^{-1} equals m_H then gave an upper bound of $m_H \approx 790 GeV$. Scaling with a Gaussian fixed point was approximately verified, with critical exponents of 0.4 (as compared with an expected 0.5). Whitmer also demonstrated that for fixed am_H the renormalized coupling λ_R is a monotone increasing function of the bare coupling λ; hence the Higgs bound can be effectively computed in the $\lambda = \infty$ theory.

Turning now to the current status of the bounding problem as reported at this conference by Weisz, Kuti, Bitar, Einhorn, Lang, Shen and Lin, we can summarize as follows:

1. Very different calculational methods are in good agreement. Weisz and Lüscher used a totally analytic method, involving matching a "hopping parameter" expansion outside the critical region with a perturbation theory calculation inside the critical region, whereas Kuti used a numerical Monte Carlo method.

2. In 1983–84 triviality of ϕ^4 was only a conjecture. While still not proved rigorously, there is now very good numerical evidence for the validity of the triviality conjecture. The numerical calculations are all consistent with a single scaling region governed by a Gaussian (free field) fixed point; the correct critical indices are obtained and there is even some evidence for the expected logarithms. There is no evidence for any non-trivial fixed points, which could lead to the existence of ϕ^4 as a strongly coupled continuum theory.

3. The $O(4)$ theory is still far from $O(2N)$, $N \to \infty$ (as a result of factors $2N+8$ in crucial places), and so large N expansion results, while interesting, do not invalidate the above statements.

4. The evaluation of m_H is still best done from $U_\Omega(\bar{\varphi})$ but the "histogram method" of determining U_Ω from a fit to $P(\bar{\varphi})$ is not accurate enough; Kuti uses a Monte Carlo method to directly compute $\partial U_\Omega/\partial\bar{\varphi}$. The bottom line is of course to determine an upper bound on m_H. Kuti quotes a bound of $550 GeV$, based on a crossover condition $m_H = 1/5a^{-1}$ (i.e., corresponding to a correlation length $m_H^{-1} = 5$ lattice units, significantly larger than used in the earlier calculations). In an alternative calculation, Bitar and Bhanot obtain a bound of $620(50) GeV$, based on a crossover condition of $m_H = a^{-1}$.

5. There was some controversy over the issue of the scheme dependence of Kuti's (or any other) upper bound calculation. For example, Bitar and Bhanot report that in discretizations on different lattices, the Higgs mass bound monotonically increases with the coordination number (number of nearest neighbors for a site) of the lattice. In this connection, I think that it is important to demand that a *natural scheme* contain no large *dimensionless numbers*, either as coefficients of irrelevant operator terms in the action introduced explicitly, or implicitly (through the use of discretizations on alternative lattices), or as coefficients used in defining the crossover point. Repeating the Monte Carlo computations in a variety of natural schemes will define a spread of reasonable upper bound estimates, and indicate the inherent theoretical uncertainty of any individual calculation. This has yet to be done in a systematic way.

I turn next to the issue of finite box size effects. When there are no zero mass particles, these behave as e^{-Lm}, with L the linear dimension of the lattice and m the mass gap in lattice units. When there are zero mass particles, such as Goldstone modes, the L-dependence instead is power law. Neuberger[4] recently proposed using this L-dependence as a way of extracting the "pion decay constant" f_π in the $O(N)$ σ-model, defined by the equation

$$\left.\begin{array}{ll} \Phi^1 = \sigma, & \Phi^\alpha = \pi^\alpha & \alpha = 2, \ldots, N-1, \\ & \langle \sigma \rangle = f_\pi Z & \\ & \left\langle 0 \left| \pi^\alpha \right| \pi^\beta \right\rangle = Z \, \delta^{\alpha\beta}, & \alpha, \beta = 2, \ldots, N. \end{array}\right\} \tag{9}$$

Neuberger's conjecture is that in a box of finite size L,

$$\left\langle \vec{\varphi} \cdot \vec{\varphi} \right\rangle_L = Z^2 \left[f_\pi^2 - \frac{(N-1)\alpha}{L^2} \right] + O(L^{-4} \log^\gamma L),$$
$$\vec{\varphi} = L^{-4} \sum_x \vec{\Phi}(x), \tag{10}$$

with α a calculable, geometry-dependent number. Although not rigorously proved, this conjecture on the L-dependence is strongly supported by computations reported at this conference by Heller, Neuberger, and Lang yielding values

$$Z \approx 1,$$
$$f_\pi^2 = 0.0588(3).$$
(11)

P. Hasenfratz, in a very interesting talk, presented a class of 4-dimensional asymptotically free non-linear σ-models, which he pointed out are analogs of the $R + R^2$ renormalizable gravitational models. He studies the action

$$S = \frac{1}{f^2} \left(C_0 S_0 + S_1 + C_2 S_2 + C_3 S_3 + C_4 S_4 + C_5 S_5 \right),$$

$$S_0 = \int Tr \left(A_\mu A_\mu \right), \quad A_\mu = U^{-1} \partial_\mu U,$$
(12)

$$S_1 = \frac{1}{2} \int \left[Tr \left(\partial_\mu A_\mu \partial_\nu A_\nu \right) + Tr \left(\partial_\nu A_\mu \partial_\nu A_\mu \right) \right],$$

$$S_{2,\ldots,5} = \text{other independent dimension} - 4 \text{ operators},$$

and finds asymptotic freedom in the dimensionless couplings f^2 and $\lambda_i = f^2 C_i, i = 2, \ldots, 5$. Some important questions for the future are:

1. When the bare dimensional coupling C_0 is zero, do the dimension -4 terms $S_{1,\ldots,5}$ lead to dimensional transmutation, with the appearance of a dimensional quantity as a renormalized coupling? If so, an S_0 term will be induced in the long wavelength effective action, even though none is present in the microscopic action. The model would then give a tractable prototype for induced gravity[5] with the idea being that scale symmetry breaking may induce an order R term in the long wavelength gravitational effective action, even when none is present in the microscopic gravitational action.

2. Are the ghosts characteristic of fourth order actions eliminated by a nonperturbative mechanism? As P. Hasenfratz pointed out, there has been considerable (unresolved) speculation and controversy about the gravitational analog of this question. It should be possible to study question (1) in perturbation theory, and both questions (1) and (2) by Monte Carlo methods, and possibly also by large N expansion.

A number of talks dealt with Monte Carlo studies of more complicated systems. In the SU_2 gauge field + Higgs model, systematics of string breaking were explored. Questions investigated in models with Yukawa couplings to fermions included the phase structure, the existence of a continuum theory with strong Yukawa couplings, and the systematics of the formation of chiral condensates. Many results in the fermion models were preliminary – the full picture of this subject is still being sketched out.

I wish to spend the rest of my time looking ahead to the future, where an important need will be greatly improved statistics on larger lattices. An important ingredient in achieving this will be improved algorithms, and as an illustration in this direction I want to discuss algorithms based on *overrelaxation*. I believe that almost everything reported at this conference could have been done faster (by a factor of 2–3 at least, perhaps more) by using overrelaxation algorithms.

The basic concept of overrelaxation comes from the numerical analysis of partial differential equations. Let $\{\phi\}$ be a collection of node variables, and let $S[\{\phi\}]$ be a multiquadratic action, that is, the dependence of S on any given node variable ϕ_k is quadratic,

$$S[\{\phi\}] = A_k (\phi_k - C_k)^2 + B_k, \quad A_k > 0, \tag{13}$$

with A_k, B_k, C_k functions of the remaining node variables $\{\phi_i, i \neq k\}$. The iterative process to solve the Euler equation $\partial S/\partial \phi_k = 0$ is an algorithm to minimize S. The simplest such algorithm, the Gauss-Seidel iteration, is the replacement

$$\phi_k \rightarrow C_k, \tag{14}$$

which brings S to its minimum with respect to the node ϕ_k, with all other nodes held fixed. This algorithm clearly gives the largest single step reduction of S, but it is not the most efficient procedure when coherent effects over the whole lattice of nodes are taken into account. One, in fact, does much better to use an overrelaxed Gauss-Seidel

$$\phi_k \rightarrow \phi'_k = \omega C_k + (1 - w)\phi_k, \tag{15}$$

with ω a parameter called the overrelaxation parameter, in which some memory of the old value ϕ_k (which contains information about long-range correlations) is retained in the update. When $\omega = 1$, Eq. (15) reduces to the Gauss-Seidel algorithm of Eq. (14). Convergence is guaranteed as long as S is monotone decreasing, which requires

$$O \leq S\left[\ldots, \phi_k\right] - S\left[\ldots, \phi_k'\right] = A_k \left(\phi_k - \phi_k'\right)^2 \left(\frac{2}{\omega} - 1\right), \tag{16}$$

and so the useful range of ω is

$$0 < \omega < 2. \tag{17}$$

A theoretical analysis[6] shows that the correlation time τ for the decay of errors in the initial guess is

$$\tau \sim L^2 \quad \text{Gauss} - -\text{Seidel}$$
$$\tau \sim L \quad \text{overrelaxed Gauss} - -\text{Seidel for } \omega = \omega_{OPT}, \tag{18}$$

with L the linear dimension of the lattice in lattice units. The optimum value ω_{OPT} of ω scales for large lattices as

$$\omega_{OPT} \approx \frac{2}{1 + \frac{C}{L}}, \tag{19}$$

with C a geometry-dependent number which is of order 3 for 4-dimensional lattices with periodic boundary conditions.

Since the heat bath Monte Carlo algorithm is the stochastic analog of the Gauss-Seidel algorithm, and since Gauss-Seidel is not optimal, it is clear that heat bath is not optimal as well. In 1981, I proposed[7] a stochastic version of the overrelaxation algorithm, described by the normalized transition probability

$$W_\omega\left[\ldots, \phi_k \to \phi_k'\right] = \left[\frac{\beta A_k}{\pi \omega(2 - \omega)}\right]^{1/2} \exp\left\{-\frac{\beta A_k}{\omega(2 - \omega)} \left[\phi_k' - \omega C_k - (1 - \omega)\phi_k\right]^2\right\}. \tag{20}$$

When $\omega = 1$, $W_1 \propto \exp\left\{-\beta S\left[\ldots, \phi_k'\right]\right\}$ and Eq. (20) reduces to heat bath, while some simple algebra shows that for all ω, W_ω satisfies the detailed-balance condition

$$\frac{W_\omega\left[\ldots, \phi_k \to \phi_k'\right]}{W_\omega\left[\ldots, \phi_k' \to \phi_k\right]} = \frac{e^{-\beta S[\ldots, \phi_k']}}{e^{-\beta S[\ldots, \phi_k]}}. \tag{21}$$

Equation (20) is equivalent to the stochastic difference equation

$$\phi_k^{n+1} - \phi_k^n = -\omega \left(\phi_k^n - C_k\right) - \left[\frac{\omega(2 - \omega)}{4\beta A_k}\right]^{1/2} \eta_{n,k},$$
$$\left\langle \eta_{n,k} \eta_{n',k'}\right\rangle_\eta = 2\delta_{n,n'} \delta_{k,k'}. \tag{22}$$

An analysis of the case of free field theory[8] shows that the Markov chain correlation time is improved by overrelaxation just as in Eq. (18).

The models of actual computational interest are not multiquadratic, but nonetheless, as I will show, overrelaxation is still useful. To illustrate this, let us consider the application of overrelaxation to the $O(N)$ Higgs model, which in the usual discretized form has the action

$$S = \text{quadratic} + \lambda \sum_x \left[\sum_a \phi_a(x)^2 - 1 \right]^2 . \tag{23}$$

Following a suggestion of Whitmer,[9] let us consider the alternative action

$$\tilde{S} = \text{quadratic} + \lambda \frac{1}{2d} \sum_x \sum_{\hat{\mu}} \left[\sum_a \phi_a(x)\phi_a(x + \hat{\mu}) - 1 \right]^2 , \tag{24}$$

where $\{\hat{\mu}\}$ are the $2d$ unit links emanating from the node at x. The action $\tilde{S}$ is equivalent to S in the continuum limit, and is multiquadratic, and so one can construct a Gaussian $\widetilde{W}_\omega$ for which any node update obeys

$$\frac{\widetilde{W}_\omega\left[\phi \to \phi'\right]}{\widetilde{W}_\omega\left[\phi' \to \phi\right]} = \frac{e^{-\tilde{S}[\phi']}}{e^{-\tilde{S}[\phi']}}. \tag{25}$$

In the unbroken symmetry phase ($m_0^2 > 0$), $\tilde{S}$ is bounded from below and is an acceptable lattice action, and hence $\widetilde{W}_\omega$ gives an acceptable Monte Carlo algorithm. In the broken symmetry phase, S is bounded from below but $\tilde{S}$ is not bounded from below for general m_0. (It is stabilized by the kinetic terms for $m_0^2 \lesssim 1/L^2$, which one could argue is a reasonable condition to impose.) However, one can get an overrelaxed algorithm based on the action S by using $\widetilde{W}_\omega$ as a *preselection* in a Metropolis algorithm with equilibrium at S. That is, one uses $\widetilde{W}_\omega\left[\phi \to \phi'\right]$ to generate a trial ϕ', which one then accepts with the conditional probability[10]

$$P_A = \text{Min}\left[1, \frac{\widetilde{W}_\omega\left[\phi \to \phi'\right]}{\widetilde{W}_\omega\left[\phi' \to \phi\right]} = \frac{e^{-\tilde{S}[\phi']}}{e^{-\tilde{S}[\phi']}}\right], \tag{26}$$

or else keeps ϕ. Substituting Eq. (25) into Eq. (26), we see that the Metropolis acceptance simplifies to

$$
\begin{aligned}
P_A &= \text{Min}\left[1, e^{\tilde{S}[\phi'] - s[\phi']} / e^{\tilde{S}[\phi] - S[\phi]}\right] \\
&= \text{Min}\left[1, 1 + O(1/L^2)\right],
\end{aligned}
\tag{27}
$$

and hence the acceptance approaches unity in the limit of fine lattices. So the Metropolis step, which shifts the equilibrium action from $\tilde{S}$ to S, is an efficient algorithm.

Some remarks are needed on the strategy for fixing ω. In solving a differential equation by overrelaxation, if we plot the decay with computational time of the errors in an initial guess, we get something like what is shown in Fig. 1. The point is that $\omega = 1$ is most effective at damping short-wavelength transients, and so initially the errors decrease fastest with $\omega = 1$; they can even *grow* initially with $\omega = \omega_{OPT}$. Eventually, the advantage of $\omega = \omega_{OPT}$ for damping long-wavelength errors appears, and the curves cross. If we wish to achieve the smallest error possible at any finite computational time, we should clearly step through a sequence of ω values, beginning at $\omega = 1$ and approaching an asymptotic value ω_{OPT}, as shown in Fig. 2. In the differential equation application, under reasonable approximations, a well-defined sequence of ω-values calculated from Chebyshev polynomials gives the optimal error decay. In practice, results almost as good are obtained by the "poor man's" version of doing 1 to 3 lattice sweeps with $\omega = 1$, and then jumping to ω_{OPT}, and it can be shown that even with only one $\omega = 1$ sweep this simplified stepping guarantees a monotone decrease of the error.[11]

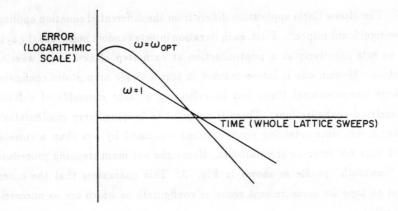

Figure 1.

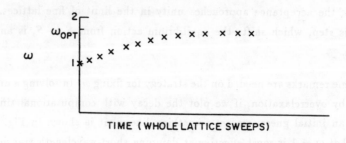

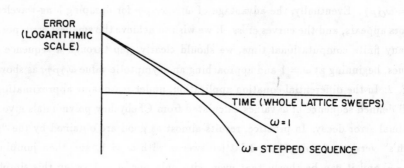

Figure 2.

The Monte Carlo application differs from the differential equation application in two significant respects. First, each iteration injects random noise into the system, and so acts effectively as a reintroduction at each step of new short wavelength structure. Second, one is interested not in just a single asymptotic configuration at large computational time, but in collecting a large ensemble of sufficiently uncorrelated configurations. There is no point in keeping every configuration for measurement, since retaining configurations separated by less than a correlation length does not improve the statistics. Hence the optimum stepping procedure [12] is a "sawtooth" profile as shown in Fig. 3. This guarantees that the ensemble stored on tape for measurement contains configurations which are as uncorrelated as possible.

The method described above for overrelaxing the $O(N)$ Higgs model has an analog for $SU(N)$ lattice gauge theory. The basic idea is that the continuum Yang-

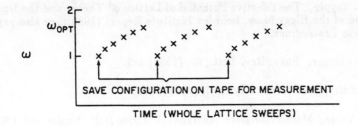

Figure 3.

Mills action *is* multiquadratic, and so one can construct a $\widetilde{W}_\omega$ which satisfies detailed balance with respect to a Cartesian discretization of the Yang-Mills action. One can then construct a Wilson lattice gauge theory analog $\widetilde{W}_\omega$, which has its equilibrium at

$$\hat{S} = S_{Wilson} + \Delta S,$$
$$\Delta S / S_{Wilson} = O(1/L^2).$$

(28)

Again, the procedure is to use $\widetilde{W}_\omega$ as a Metropolis preselection with equilibrium at S_{Wilson}, giving a Metropolis acceptance of $1 + O(1/L^2)$. The construction sketched above has been described in detail elsewhere,[13] and is currently being tested by Gyan Bhanot.

To conclude: better algorithms, which deal with critical slowing down, may soon be available. In particular, overrelaxation algorithms improve the correlation time while requiring little or no more computation and memory. The systematics of their behavior should be understood in the course of the next few months.

This work was supported by the Department of Energy under Grant No. DE-AC02 -76ERO2220.

REFERENCES

1. R. Dashen and H. Neuberger, Phys. Rev. Lett., 50, (1983) 1897.

2. C. Whitmer, Monte Carlo Methods in Quantum Field Theory, unpublished Princeton University dissertation (1984).

12

3. M.M. Tsypin, The Effective Potential of Lattice ϕ^4 Theory and the Upper Bound of the Higgs Mass, Lebedev Institute Report (1985); see also paper in these Proceedings.

4. H. Neuberger, Phys. Rev. Lett., 60, (1988) 889.

5. S.L. Adler, Rev. Mod. Phys., 54, (1982) 729.

6. R.S. Varga, Matrix Iterative Analysis (Prentice-Hall, Englewood Cliffs, 1962) chapters 3 and 4.

7. S.L. Adler, Phys. Rev., D23, (1981) 2901.

8. J. Goodman and A.d. Sokal, Phys. Rev. Lett., 56, (1986) 1015.
S.L. Adler, Phys. Rev., D37, (1988) 458.
H. Neuberger, Phys. Rev. Lett., 59, (1987) 1877.

9. C. Whitmer, Phys. Rev., D29, (1984) 306 and Ref. [2].

10. S. Gottlieb, W. Liu, D. Toussaint, and R.L. Sugar, Phys. Rev., D35, (1987) 2611 for a Metropolis application similar in spirit to the one made here.

11. R.S. Varga, See Chapter 5 in Ref. [6] for a detailed discussion.

12. S.L. Adler, Phys. Rev., D37, (1988) 458 (See note added for the discussion of ω-stepping.).

13. Ref. [12] and S.L. Adler, Metropolis Overrelaxation for Lattice Gauge Theory for General Relaxation Parameter ω, Phys. Rev. D (in press).

CHIRAL GAUGE THEORY ON A LATTICE

Sinya Aoki

Physics Department
Brookhaven National Laboratory
Upton, NY 11973, USA

ABSTRACT

We discuss the quantization of chiral gauge theories by lattice regularization, carefully treating the effects of the chiral anomaly. We derive a chiral gauge invariant lattice fermion action from a chiral gauge variant Wilson fermion action without changing its partition function. By lattice power counting for this formula we show that anomalous gauge theories as well as anomaly-free gauge theories are renormalizable even in 4-dimensions. Some applications and implications of this result and problems therein are discussed.

1. PROBLEM OF CHIRAL GAUGE THEORY

1.1. Motivation

Lattice regularization[1] provides us with a powerful tool to investigate quantum field theories, such as QCD or the gauge-Higgs system. However, successes of lattice regularization have been limited to purely bosonic systems or to systems with fermion-gauge field vector-like interactions. The reason for this is that we can not define the lattice fermion coupled to gauge field in a chirally invariant way.[2] This fact is closely related to the local gauge anomaly.[3] Because the lattice regularization is well-defined, an anomalous symmetry cannot be maintained on a lattice. In the continuum case, the anomalous symmetry is broken by the regularization, for example, Pauli-Villars regularization or dimensional regularization.

From this observation, it seems possible to construct a chirally gauge invariant action for the anomaly-free case. However, it is not so easy. There are two ways of breaking chiral gauge symmetry: one is by the anomaly, the other is the breaking which can be written as gauge variant local terms. Anomaly-free means that a pure anomaly is canceled out among all fermions contributions; however, it does not mean the cancellation of the gauge variant local terms. We should add by hand the gauge variant local counterterms order by order to recover the gauge invariance of the final results. Furthermore the cancellation of a pure anomaly is only true for the infinite cut-off limit. Because of the above two reasons, we cannot generally have chiral gauge invariant action with finite cut-off even for the anomaly-free case.

In this talk we will discuss the quantization of anomalous chiral gauge theories as well as anomaly-free theories by using lattice regularization. One important physical motivation for this work is to put the full Weinberg-Salam model (with fermions) on a lattice in order to investigate it by non-perturbative methods such as strong coupling expansion or Monte-Carlo simulation. The other theoretical motivation is to investigate the consistency; for example, renormalizability or unitarity, of anomalous gauge theories. In this talk we concentrate on this point and show that anomalous gauge theories are renormalizable in the sense of lattice power counting.[4]

1.2. Quantization of anomalous gauge theories in the continuum approach

First we briefly sketch how difficult is the quantization of anomalous gauge theories in the continuum approach by following a paper by Gross and Jackiw.[5]

We denote the action by $S(q)$, where q represents any field involved in the theory. In this short-hand notation a gauge transformation is written as

$$q'(x) = q(x)^{\omega(x)}, \qquad \omega(x) = \exp[i\theta(x)]$$

where $\omega(x) \in U(N), SU(N)$ or other Lie groups, is a gauge function. For example, $q'(x)$ is explicitly given by

$$\psi'(x) = [\omega(x)P_L + P_R]\psi(x)$$
$$\bar{\psi}'(x) = \bar{\psi}(x)[\omega^{-1}(x)P_R + P_L]$$
$$A'_\mu(x) = \omega(x)A_\mu(x)\omega^{-1}(x) + \frac{1}{g}\omega^{-1}(x)\partial_\mu\omega(x)\,.$$

In order to make a perturbative expansion (more precisely, to get a propagator for a gauge field) we have to fix the gauge. That is achieved by inserting *unity* :

$$\prod_x \int d\omega(x)\exp[S_{GF}(q^{\omega^{-1}})] = 1$$

into the partition function. Here the Faddeev-Popov determinant[6] is included in the gauge fixing function S_{GF}. Then the partition function becomes

$$Z = \int \mathcal{D}q \, \exp[S(q)] \int \mathcal{D}\omega \, \exp[S_{GF}(q^{\omega^{-1}})]$$

$$= \int \mathcal{D}q^\omega \, \mathcal{D}\omega \, \exp[S(q^\omega) + S_{GF}(q)]$$

$$= \int \mathcal{D}q \, \mathcal{D}\omega \, \exp[S_{\text{eff}}(q,\omega) + S_{GF}(q)],$$

where we define

$$\mathcal{D}q \, \exp[S_{\text{eff}}(q,\omega)] \equiv \mathcal{D}q^\omega \, \exp[S(q^\omega)], \qquad S_{\text{eff}}(q,1) = S(q).$$

Since the theory is anomalous,

$$S_{\text{eff}}(q,\omega) = S_{\text{eff}}(q,1) + \int dx \, \theta(x)\mathcal{A}(x) + O(\theta^2)$$

for infinitesimally small θ, where $\mathcal{A}(x)$ is the so-called non-Abelian anomaly.[3] For finite θ,

$$S^{WZ}(q,\omega) \equiv S_{\text{eff}}(q,\omega) - S_{\text{eff}}(q,1)$$

is nothing but the Wess-Zumino term.[7] The field ω is called the Wess-Zumino scalar. Finally we get

$$Z = \int \mathcal{D}q \, \mathcal{D}\omega \, \exp[S(q) + S^{WZ}(q,\omega) + S_{GF}(q)].$$

This form was first obtained by Harada-Tsutsui[8] and it is equivalent to the proposal by Fadeev-Shatashivili[9] who add the Wess-Zumino term to the original action by hand to quantize the anomalous gauge theory. We have to integrate both q and ω interacting through the Wess-Zumino term to quantize anomalous gauge theory correctly.

It is easy to see that $S(q) + S^{WZ}(q,\omega)$ has gauge invariance such that

$$q'(x) = q^h(x), \qquad \omega'(x) = h^{-1}\omega(x).$$

From this property arises a hope that the theory defined by $S(q) + S^{WZ}(q,\omega)$ may be renormalizable. However, since $S^{WZ}(q,\omega)$ is highly non-linear in 4-dimensions, it is difficult to work with it.

2. LATTICE QUANTIZATION

Next let us apply the above procedure to the lattice action. From the observation in the continuum case, we should take a gauge *variant* action at the begining in order to quantize an anomalous gauge theory. Therefore we take the Wilson fermion action[10] for the fermionic part of the action, which is given by

$$S_F = \int \bar{\psi}\gamma_\mu(D_\mu P_L + \partial_\mu P_R)\psi + ar\bar{\psi}\partial^2\psi$$

$$= a^d \sum_{n,\mu} \frac{1}{2a}\bar{\psi}_n[(U_{n,\mu}P_L + P_R)\psi_{n+\mu} - (U^\dagger_{n-\mu,\mu}P_L + P_R)\psi_{n-\mu}]$$

$$- a^d \sum_{n,\mu} \frac{r}{2a}\bar{\psi}_n(\psi_{n+\mu} + \psi_{n-\mu} - 2\psi_n)$$

$$\equiv S_{\text{inv}}(\varphi) + S_W(\varphi).$$

Here φ represents $U_{n,\mu}, \bar{\psi}$ and ψ , S_{inv} is a gauge invariant part of S_F and S_W is a gauge variant part of S_F, that is, the Wilson term. The total action is $S_F + S_G$ where S_G is the pure gauge action. We call a field φ of this original action the physical field and denote it by φ^p, if necessary. A gauge transformation is given by $\varphi'_n = \varphi^g_n$ and explicitly written as

$$\psi'_n = (g_n P_L + P_R)\psi_n$$

$$\bar{\psi}'_n = \bar{\psi}_n(g^\dagger_n P_R + P_L)$$

$$U'_{n,\mu} = g_n U_{n,\mu} g^\dagger_{n+\mu}.$$

As in the continuum case, we can insert unity :

$$\prod_n \int dg_n \exp S_{GF}(\varphi^{g^{-1}}) = 1$$

into the partition function to fix the gauge in the perturbative expansion. However, for non-perturbative calculations, it is not necessary to fix the gauge on the lattice. In that case we can take $S_{GF} = 0$ since $\int dg_n = 1$. This is possible only on a lattice. By inserting the gauge fixing condition into the partition function and making a change of integration variables such as $\varphi^r_n = \varphi^{g^{-1}_n}_n$ where we call this new field a renormalizable field and attach the suffix r, we get

$$Z = \int \mathcal{D}g\,\mathcal{D}\varphi \; \exp[S_{\text{inv}}(\varphi^g) + S_W(\varphi^g) + S_{GF}(\varphi)]$$

$$= \int \mathcal{D}g\,\mathcal{D}\varphi \; \exp[S_{\text{inv}}(\varphi) + S_W(\varphi^g) + S_{GF}(\varphi)],$$

where the suffix r is omitted and

$$S_W(\varphi) = a^d \sum_{n,\mu} \frac{r}{2a} [\bar{\psi}_n g_n^\dagger P_R(\psi_{n+\mu} + \psi_{n-\mu} - 2\psi_n)$$
$$+ (\bar{\psi}_{n+\mu} + \bar{\psi}_{n-\mu} - 2\bar{\psi}_n) g_n P_L \psi_n] .$$

This term represents the violation of gauge invariance and includes all information concerning the anomaly. For example the non-Abelian anomaly can be obtained by this term in lattice perturbation theory.[11]

It is noted that $S_{\text{inv}}(\varphi) + S_W(\varphi^g)$ also is gauge invariant under

$$\varphi_n' = \varphi_n^{h_n}, \qquad g_n' = h_n^{-1} g_n .$$

Because the above action is bilinear for fermions, it is possible to do Monte-Carlo simulation with this action. This form[4,12] is also suitable for a perturbative calculation of the 2-dimensional chiral Schwinger model.[13]

Since there are no couplings among g_n in the action, we can integrate the g_n field[4] explicitly. We define

$$\exp[K(\varphi)] = \prod_n \int dg_n \ \exp[S_W(\varphi^g)] ;$$

then we get, for example, for $g_n \in U(1)$,

$$\exp[K(\varphi)] = \prod_n I_0(2\sqrt{\bar{A}_n A_n})$$

$$\bar{A}_n = -a^d \frac{ar}{2} \sum_\mu \bar{\psi}_n P_R \partial_\mu^2 \psi_n$$

$$A_n = -a^d \frac{ar}{2} \sum_\mu (\partial_\mu^2 \bar{\psi}_n) P_L \psi_n .$$

Furthermore, because of the Grassmann property of fermions, $K(\varphi)$ is a finite polynomial of $\bar{\psi}$ and ψ. For example, for $g_n \in U(1)$ in 2 dimensions, it is given by

$$K(\varphi) = a^{2d} \frac{a^2 r^2}{4} \sum_n \sum_{\mu,\nu} \bar{\psi}_n P_R \partial_\mu^2 \psi_n \times (\partial_\nu^2 \bar{\psi}_n) P_L \psi_n ,$$

and for $g_n \in U(1)$ in 4-dimensions, it is given by

$$K(\varphi) = a^{2d} \frac{a^2 r^2}{4} \sum_n \sum_{\mu,\nu} \bar{\psi}_n P_R \partial_\mu^2 \psi_n \times (\partial_\nu^2 \bar{\psi}_n) P_L \psi_n$$

$$- \frac{1}{4} a^{4d} \sum_n [\frac{a^2 r^2}{4} \sum_{\mu,\nu} \bar{\psi}_n P_R \partial_\mu^2 \psi_n \times (\partial_\nu^2 \bar{\psi}_n) P_L \psi_n]^2 .$$

It is noted that $K(\varphi)$ for general g_n is also local and gauge invariant but non-linear.

3. RENORMALIZABILITY AND PHYSICAL INTERPRETATION

From lattice power counting,[14] which means that we count the order of lattice spacing a of the given diagram, for the non-linear action $S_{inv}(\varphi) + K(\varphi)$, we have shown[4] that the superficial degree of divergence D of a given Feynman diagram in four dimensions is

$$D = 4 - (E_G + E_g) - \frac{3}{2}E_f \, ,$$

where E_G, E_g and E_f are the numbers of external lines for gauge field, ghost field, and fermion, respectively. This shows that anomalous gauge theories as well as anomaly-free theories are renormalizable even in four dimensions. The effect of the anomaly induced by the g integral is renormalizable. If we add the kinetic term for the g field to the action before the g integration, we can identify g as a polar part of the Higgs field.[15] However, this makes the g integration impossible and the theory un-renormalizable.

There are some remarks.

1) Though the physical field φ^p does not have species doubling modes, the gauge invariant, non-linear action for the renormalizable field φ^r may have doubling problems. However, it is stressed that we do not attempt to propose the chiral gauge invariant lattice action for the physical fields though Nielsen-Ninomiya's theorem[2] does not apply to the non-linear action. We introduce φ^r to show the renormalizability of the theory and the physical interpretation should be made for φ^p. This relation between φ^r and φ^p reminds us of the relation between the renormalizable gauge and the unitary gauge in the gauge-Higgs system.

2) Since the action for φ^p is gauge variant, there exist a non-trivial relation between φ^p and φ^r. For example, to see the relation between fermion fields with $g_n \in U(1)$, we introduce a generating function for fermion fields such as

$$Z(\eta, \bar{\eta}) = \int \mathcal{D}\bar{\psi}\,\mathcal{D}\psi\,\mathcal{D}U \ \exp[S_{inv} + S_W + \bar{\eta}\psi^p + \bar{\psi}^p\eta]$$

$$= \int \mathcal{D}\bar{\psi}\,\mathcal{D}\psi\,\mathcal{D}U \ \exp[S_{inv} + \bar{\eta}P_R\psi^r + \bar{\psi}^r P_L\eta]\prod_n I_0(2\sqrt{\bar{D}_n D_n}) \, ,$$

where

$$\bar{D}_n = \bar{A}_n + \bar{\psi}_n^r P_R \eta_n, \quad D_n = A_n + \bar{\eta}_n P_L \psi_n^r \, .$$

From the above identity it is easy to get equations such that

$$< (P_R \psi_n^p)^\alpha (\bar{\psi}_m^p P_L)^\beta > = < (P_R \psi_n^r)^\alpha (\bar{\psi}_m^r P_L)^\beta >$$
$$< (P_L \psi_n^p)^\alpha (\bar{\psi}_m^p P_L)^\beta > = < (P_L \psi_n^r)^\alpha (\bar{\psi}_m^r P_L)^\beta A_n >$$
$$< (P_L \psi_n^p)^\alpha (\bar{\psi}_m^p P_R)^\beta > = < (P_L \psi_n^r)^\alpha (\bar{\psi}_m^r P_R)^\beta A_n \bar{A}_m >$$
$$+ \delta_{nm} < (P_L \psi_n^r)^\alpha (\bar{\psi}_n^r P_R)^\beta >$$

in 2-dimensions. Similar but more complicated equations can also be obtained in 4-dimensions. By using these equation we can calculate all vacuum expectation values of the physical field φ^p if we know all vacuum expectation values of the renormalized field φ^r.

3) If we interpret $\bar{\psi}^p$, ψ^p and A_μ^p as real physical fields, we can observe interesting properties as follows:

- The physical fermion field is not confined because gauge invariance does not hold for the physical fields.
- Mass generation for the gauge field by quantum effect is allowed because of the same reason.

The above two properties may correspond to the fact that the weak gauge bosons have mass and the leptons are not confined.

4) It is noted that from the renormalizable theory, which has a finite number of counterterms, by some non-linear transformation we may construct the theory which seems to be un-renormalizable because the number of counter terms is infinite. However, in this case the number of independent counter terms is finite. The non-linear sigma model in 4-dimensions may be one example of this type of theory.[16] It is interesting to see whether the anomalous theory is of this type or not.

We are now at the starting point to investigate an (anomalous) chiral gauge theory on a lattice; we should carefully study both perturbative and non-perturbative properties of this theory. As a first step, the perturbative expansion for a non-linear action is now under investigation. After getting some information from that calculation we would like to do Monte-Carlo simulations in the near future.

Acknowledgements

I would like to thank Prof. T. L. Trueman for careful reading of the manuscript and useful comment. I also thank Prof. M. Creutz, Prof. R. Shrock and Dr. I-Hsiu Lee for useful discussion. This work was supported by the U.S. Department of Energy under contract DE-AC02-76CH00016.

REFERENCES

1. K. G .Wilson, Phys. Rev. D10, 2445 (1974).

2. H. B. Nielsen and M. Ninomiya, Nucl. Phys. B185, 20 (1981); B193, 173 (1981).

3. W. A. Bardeen, Phys. Rev. 184, 1828 (1969).

4. S. Aoki, Phys. Rev. Lett. 60, 2109, (1988); Phys. Rev. D37, (to be published in 1988).

5. D. J. Gross and R. Jackiw, Phys. Rev. D6, 477 (1972).

6. L. D. Faddeev and V. Popov, Phys. Lett. 25B, 29 (1967).

7. J. Wess and B. Zumino, Phys. Lett. 37B, 95 (1971).

8. K. Harada and I. Tsutsui, Phys. Lett. 183B, 311 (1987).

9. L. D. Faddeev and S. L. Shatashvili, Phys. Lett 167B, 225 (1986).

10. K. G. Wilson, 'Quark and Strings on a Lattice.' in: New Phenomina in Subnuclear Physics, ed. A. Zichichi (Plenum, New York, 1977).

11. A. Coste, C. Korthals-Altes and O. Napoly, Phys. Lett. 179b, 125 (1986); S. Aoki, Phys. Rev. D35, 1435 (1987).

12. K. Funakubo and T. Kashiwa, Phys. Rev. Lett. 60, 2113 (1988).

13. R. Jackiw and R. Rajaraman, Phys. Rev. Lett. 54, 1219 (1985).

14. H. Kawai, R. Nakayama and K. Seo, Nucl. Phys. B189, 40 (1981).

15. L. H. Karsten, 'The lattice Fermion Problem and Weak Coupling Perturbation Theory.' in: Field Theoretical Method in Particle Physics, ed. W. Rühl (Plenum, New York, 1980); J. Smit, Acta Physinica Polonica B17, 531 (1986); P. V. D. Swift, Phys. Lett. 145B, 256 (1984).

16. P. Hasenfratz, in; these proceedings.

LATTICE HIGGS MASS BOUNDS AND DIFFERENT CUTOFF SCHEMES

Khalil Bitar and Gyan Bhanot

Supercomputer Computations Research Institute
Florida State University
Tallahassee, FL 32306

ABSTRACT

The aim of this work is to study the regularization dependance of the lattice Higgs mass bound. For this purpose, we simulated the Higgs sector of the standard Weak-Electromagnetic theory at $\lambda = \infty$ on three types of lattices; a) hyper-cubic with interactions along the axes, b) hyper-cubic with interactions of equal strength along the axes and along plane diagonals and c) hyper-triangular lattices with interactions along the axes. For each of these lattice types, we obtained the effective potential from which we compute f_π and λ_r as a function of the scalar (Higgs) mass M_H. Both these give a bound $M_{phys}^H < 620(50)$ Gev where the error comes from the variation in our results from changing the lattice type/interaction. Since we also find that the mass bound increases systematically as the number of nearest neighbor terms in the interaction increases, a reliability estimate on such bounds (obtained near $M_H = 1$) may not be meaningful.

Experimentally, the single missing ingredient in the standard model of the weak and electromagnetic interactions is the Higgs particle. The standard model assumes that the Higgs sector consists of an $O(4)$ invariant fundamental scalar field. It is quite possible that the Higgs is actually composite. This would mean that when one reaches energies higher than the binding energies of the composite Higgs particles, new physics will be revealed. It is widely believed that within the standard model, one can estimate the energy before which either the Higgs particle must be found or else new forces (for example those between the constituents of the Higgs particle if it is composite) must come into play.

The argument that one might get a bound on the Higgs mass in the standard model was constructed by Dashen and Neuberger [1] and numerical studies were first presented by Whitmer and Tsypin [2]. We will present the argument here in a slightly different way from Ref. [1] so as to focus on the issues relevant to our discussion.

Consider the Higgs sector of the Glashow, Salam, Weinberg model with the gauge coupling temporarily turned off. The continuum theory is defined by the Lagrangian density,

$$L = \frac{1}{2}\partial_\mu \vec{\phi}(x)\partial_\mu \vec{\phi}(x) + V(|\vec{\phi}(x)|) \tag{1}$$

where,

$$V(|\vec{\phi}|) = -\frac{1}{2}m_0^2|\vec{\phi}|^2 + \lambda(|\vec{\phi}|^2)^2 \tag{2}$$

and,

$$\vec{\phi} = (\phi_1, \phi_2, \phi_3, \phi_4), \quad \phi_i \epsilon(-\infty, \infty) \tag{3}$$

We will work exclusively in the broken phase (ie. m_0^2 is positive).

The operator $|\vec{\phi}| = \sqrt{\phi_1^2 + \phi_2^2 + \phi_3^2 + \phi_4^2}$ has a non-zero vacuum expectation value $v = \frac{m_0}{2\sqrt{\lambda}}$. Shifting $|\vec{\phi}|$ by this amount one finds that the theory now describes a triplet of massless Goldstone Bosons plus a massive scalar (Higgs) particle with tree level mass $M_{phys}^H = \sqrt{2}m_0$.

One would like to define a regularization prescription that allows a non-perturbative description of the mass and coupling constant renormalization. We will regularize the theory by defining it on a lattice. On the lattice also, there are two bare parameters, the quartic coupling λ and the mass parameter m_0. The lattice theory has action given by,

$$S = \frac{1}{2}\frac{8}{N_c}\sum_{i,\hat{\mu}}|\vec{\phi}_i - \vec{\phi}_{i+\hat{\mu}}|^2 - \frac{m_0^2}{2}\sum_i|\vec{\phi}_i|^2 + \lambda\sum_i(|\vec{\phi}_i|^2)^2 \tag{4}$$

with the $\vec{\phi}_i$ fields defined on the lattice sites i. The sum over $\hat{\mu}$ is over all forward directed links connected to the sites. N_c is the coordination number, or the number of links connected to each site (counting both forward and backward links). For example, for a cubic lattice with interactions along the axes, $N_c = 8$. N_c is included to generalize the discussion to other lattice types and extra interactions. It is easy to see that the continuum limit of the theory of Eq. 4 has Lagrangian density given by Eq. 1.

Let us define the renormalization prescription we will use. Consider the theory for a fixed value of λ. One can vary m_0 to obtain any given value for the renormalized

Higgs mass M_H (in dimensionless lattice units) and then compute the renormalized quartic coupling λ_r. M_H is related to the physical mass by $M_H = \frac{M^H_{phys}}{\Lambda}$ where Λ is the cut-off parameter. The continuum limit is obtained by letting $\Lambda \to \infty$ whilst keeping M^H_{phys} fixed. There is a great deal of evidence [3,4,6] that in this limit, the theory is trivial; ie. $\lambda_r \to 0$ as $\Lambda \to \infty$. Conversely, as Λ is decreased from very large values, λ_r increases.

Now the theory only makes sense as a cutoff theory when $M_H < 1$ since the cut-off can never be smaller than the physical mass. Hence, for any fixed λ, one can compute the renormalized quartic coupling λ_r^{max} at which $M_H = 1$. This is the maximum value of the renormalized coupling for which the theory makes sense. One can then ask how this maximum renormalized coupling changes as the bare λ is changed. It seems reasonable to expect that for a fixed value of M_H, the largest λ_r will be obtained by starting from a bare $\lambda = \infty$. A variety of numerical [2,3,4] and analytical [6,7] studies indicate that this expectation is indeed true. Since we are only interested in this maximum possible value of λ_r^{max}, we will restrict our study to the $\lambda = \infty$ theory.

Consider now what happens when the gauge coupling g is turned on. Obviously, since the gauge coupling is experimentally known to be small, we can discuss what happens perturbativly in g. When a small g is turned on, the first effect will be that the three degrees of freedom associated with the Goldstone modes of the Higgs sector will become longitudinal degrees of freedom of the gauge particles through the Higgs mechanism. In other words, the gauge Bosons will acquire mass and the Goldstone particles will disappear from the spectrum. To lowest order in the gauge coupling constant, the mass M_W of the $W^\pm$ or Z^0 (we neglect the electromagnetic $U(1)$ symmetry as that is irrelevant to our discussion here) is given by,

$$M_W^2 = \frac{g^2}{32\lambda_r}(M^H_{phys})^2 \qquad (5)$$

In Eq. 5, M_W and g^2 are known from experiment. If one has an upper bound on λ_r, one has an upper bound on the Higgs mass M_H.

Another quantity that one could compute on the lattice is f_π [1,5] whose value is known from its connection to the Fermi coupling: $f_\pi = (\sqrt{2}G_F)^{-1/2} = 246$ Gev. If one measures f_π as a function of M_H then one can get an estimate of the limit of the physical region ($M_H < 1$) directly. As has been emphasized in Ref. [1,5]

the estimate via f_π depends neither on λ_r nor on the perturbative relation in g connecting M_H and M_W.

Both methods however rely on extracting a bound from where a lattice mass scale (or correlation length) is of order unity. One expects that such estimates might be strongly contaminated by cutoff effects, especially λ_r which needs both a regularization scheme and a renormalization prescription to be defined.

In the limit $\lambda = \infty$ (keeping $\frac{m_0^2}{\lambda}$ fixed), the lattice action is given by,

$$S(\vec{\phi}) = -\beta \sum_{i,\hat{\mu}} \vec{\phi}_i \cdot \vec{\phi}_{i+\hat{\mu}} \tag{6}$$

where $|\vec{\phi}| = 1$ and where β is related to the parameters λ and m_0 by

$$\beta = \frac{m_0^2}{N_c \lambda} \tag{7}$$

To find the upper bound on λ_r, we will use the effective potential method of Fukuda and Kyriakopoulos [8] as first applied to this problem by Whitmer and Tsypin [2]. Our aim is to study the effect of changing the regularization procedure on parameters in the effective potential. The effective potential is defined by putting an appropriate delta function into the path integral. Following Refs. [2,8],

$$P(\phi_c) = \phi_c^3 e^{-V S_{eff}(\phi_c)} = \int d\phi e^{-S(\vec{\phi})} \delta(\phi_c - |\frac{1}{V} \sum_i \vec{\phi}_i|) \tag{8}$$

In a practical simulation, implementing Eq. 8 via Monte Carlo amounts to generating configurations distributed according to the Boltzmann weight e^{-S} and making a histogram of the values obtained for ϕ_c in the equilibrium, thermalized distribution from independent lattice samples. We used a version of the heat-bath algorithm of Kennedy et al [9] to generate configurations and simulated the theory on lattices of linear size 4, 6, 8 and 10 with periodic boundary conditions.

A lattice type is defined in terms of the unit vectors $\vec{e}_\mu$ from which it is constructed. We worked on two types of lattices. The first was hyper-cubic with unit vectors given by

$$e_\mu^i = \delta_{i,\mu}, \quad i = 1, 2, 3, 4. \tag{9}$$

Here i labels the components of the vector. The second lattice we simulated had unit vectors defined by the condition $\vec{e}_\mu \cdot \vec{e}_\nu = \frac{1}{2}$ for $\mu \neq \nu$. One possible representation of these vectors is,

$$e_1 = (1, 0, 0, 0)$$

$$e_2 = (\frac{1}{2}, \frac{\sqrt{3}}{2}, 0, 0)$$

$$e_3 = (\frac{1}{2}, \frac{1}{2\sqrt{3}}, \frac{\sqrt{2}}{\sqrt{3}}, 0) \tag{10}$$

$$e_4 = (\frac{1}{2}, \frac{1}{2\sqrt{3}}, \frac{1}{2\sqrt{6}}, \frac{\sqrt{5}}{2\sqrt{2}})$$

It can be shown that these unit vectors result in a uniform lattice with coordination number N_c equal to twenty. The twenty neighbors connected to the site are along the directions,

$$\pm \vec{e}_\mu, \quad \text{and} \quad \pm (\vec{e}_\mu - \vec{e}_\nu), \mu > \nu \tag{11}$$

with the e_μ in Eq. 11 given by Eq.10. This lattice can be called hyper-triangular. We will denote it as lattice type T and the cubic lattice will be denoted C. Note that one can think of simulations on T as simulations on a cubic lattice with interactions of equal strength along unit vectors in the directions of Eq. 11 where the e_μ's are those of Eq. 9.

We have simulated the action of Eq. 6 on T and C. In addition, we have also simulated the theory on C with interactions of equal strength along unit vectors in the direction $\pm e_\mu$ and $\pm (e_\mu \pm e_\nu), \mu > \nu$ where again the e_μ's are those of Eq. 9. N_c for this last type of lattice (which we will call C') is 32. Figure 1 shows in two dimensions the progression from lattice type C to T to C'.

For all these simulations, we computed the probability density $P(\phi_c)$ of Eq. 8 for different values of β in the region where the renormalized Higgs mass was of order unity. The $P(\phi_c)$'s obtained were fitted to,

$$P(\phi_c) \sim \phi_c^3 e^{-V S_{eff}(\phi_c)} = \phi_c^3 \exp\{-V\lambda_r[\phi_c^2 - \eta^2]^2\} \tag{12}$$

and the parameters λ_r and η^2 obtained by a least squared fit. Figure 2 shows typical histograms we obtained for $P(\phi_c)$ and the fit to Eq. 12. The fits we obtained had an excellent χ^2 which we take to indicate that the form of Eq. 12 is a good one for the fits.

Note that after shifting the field by η, the renormalized mass is given by $M_H^2 = 8\lambda_r \eta^2$ and so can be extracted from the fit to Eq. 12. Our updating algorithm does an almost heat-bath update with an acceptance rate approaching 100 percent. The sites were chosen for updating randomly. We typically discarded 5000 sweeps.

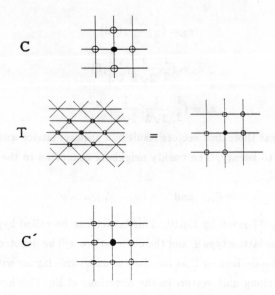

Figure 1. A two dimensional illustration of the lattice/interaction types we used. The variable at the site represented by a filled dot interacts with all the variables at sites shown as open circles. The lattice type C is simple cubic with nearest neighbor interactions along axes. The lattice type T is trianbgular with nearest neighbor interactions. T can also be interpreted as a cubic lattice with interactions of equal strength along the axes and along two opposite planar diagonals. lattice type C' is simple cubic with interactions of equal strength along axes and plane diagonals.

After that, about 10,000 sweeps were done with the histogram for $P(\phi_c)$ updated after each sweep. Our program was written for the CYBER-205 and the ETA[10]. The vectorization was done by the brute force method of having independent systems as the vector index. We typically had vector lengths of between 50 and 300. This means that although we generate only about 10,000 sweeps at each β value, the data sample is bigger by a factor equal to the vector length. Note that since we are interested in the physics when the correlation length is of order unity (and not where it is large), we do not have any problems with large auto correlation times. We used approximately 250 hours of ETA[10] time on the Florida State University supercomputer and another 200 hours on the Florida State University and Princeton CYBER-205's.

We also tried a variety of other fits with more parameters. In every instance,

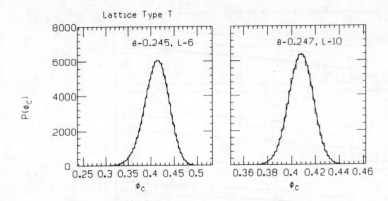

Figure 2. Two typical histogrammed distributions for $P(\phi_c)$ for lattice type T. The solid lines are a least squared fit to Eq. 12.

the χ^2 value of the fit deteriorated appreciably. However, in all the fits, we always found that the M_H value, determined from the coefficient of the quadratic term of S_{eff}, was very stable as was the position (η^2) of the minimum of the effective potential. However, to be able to compute λ_r reliably from fits to the effective potential, we found it was necessary to use the form of Eq. 12.

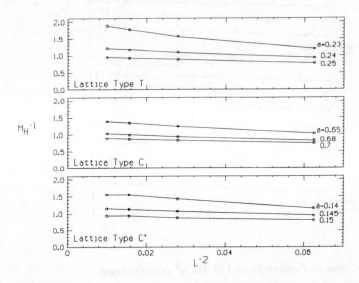

Figure 3a. $\frac{1}{M_H}$ versus L^{-2} at some typical values of β we simulated for the various lattice types. The lines are not fits.

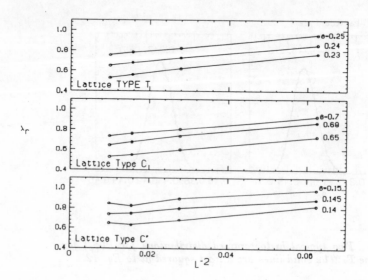

Figure 3b. *Same as in Figure 3a but now λ_r is plotted instead of $\frac{1}{M_H}$.*

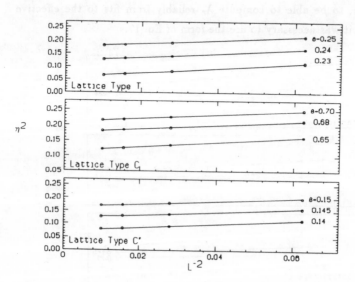

Figure 3c. *Same as Figures 3a and 3b but η^2 is now shown.*

All our data had to be extrapolated to $L = \infty$. There is a conjecture due to Neuberger [5] that $< \phi_c^2 >$ has a leading correction of $O(\frac{1}{L^2})$. We find from our results that this is true not only for $< \phi_c^2 >$ but also for η^2, M_H and λ_r. Figure 3 shows $\frac{1}{M_H}$, η^2 and λ_r plotted against $\frac{1}{L^2}$ for the various lattice types at some of the β values we simulated. The extrapolation to $L = \infty$ was done by making straight line least squared fits to data such as is shown in Figure 3.

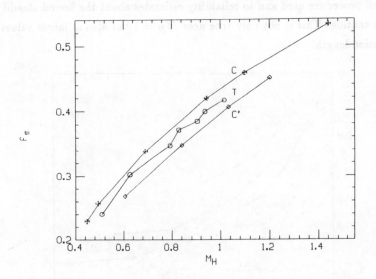

Figure 4. f_π as a function of M_H from the infinite volume extrapolation of η^2 (such as in Fig. 1). Note the systematic decrease in f_π at fixed M_H as the coordination number of the lattice increases.

As explained in [1,5], as $L \to \infty$, $< \phi_c^2 > \to Z^2 f_\pi^2$. Since there is numerical evidence that Z is close to unity for the region where $M_H \sim 1$, we will neglect Z henceforth. Further, as $< \phi_c^2 > = \eta^2$ in the infinite volume limit and our results for η^2 were very reliable, we extrapolated η^2 to find f_π^2. We emphasize here that our results for M_H and η^2 (from which we measure f_π) are not dependent on the form of Eq.12. The numbers we obtained for these quantities with other fits were almost unchanged. Figure 4 shows our results for f_π versus M_H. The physical region corresponds to $M_H < 1$ or $f_\pi < 0.40(0.03)$. Using the phenomenological value of f_π, this gives the bound $M_{phys}^H < 620(50)$. Written in terms of the W mass, this bound is $\frac{M_{phys}^H}{M_W} < 7.6(0.4)$. The error quoted here is the scatter in the bound between the three cutoff schemes we use.

Notice however that there is a systematic effect as the lattice coordination number is increased. This is evident from Figure 4 because f_π at $M_H = 1$ progressively decreases as one goes from lattice type C to T to C'. This we take as an indication that the f_π obtained from the effective potential is not independent of the cutoff procedure used and so reliability estimates about the bound should be made with caution. This is not only true near $M_H = 1$ but also at larger values of the correlation length.

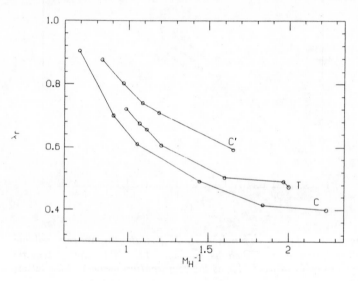

Figure 5. λ_r versus $\frac{1}{M_H}$ from the infinite volume extrapolation. Note again the small but systematic increase in λ_r at fixed M_H as the coordination number of the lattice increases.

We have also measured λ_r using fits to Eq. 12. Figure 5 shows our results for λ_r versus $\frac{1}{M_H}$ from the extrapolation to infinite volume. If one computes the bound from $M_H = 1$ as before, one gets a number consistent with the estimate from f_π. This indicates that, although one needs the form of Eq. 12 to get a stable fit for λ_r, our numbers for λ_r are reliable. Furthermore, it is known that λ_r defined via the effective potential has an infrared problem because of the two 'Goldstone' cut. This may explain the disparity between the various cutoff schemes in our numbers for λ_r at small M_H in Figure 5.

Notice that again in Figure 5, there is the same systematic effect of the bound increasing as the coordination number increases. This, coupled with the uncertainty about what the best regularization scheme is, whether the bound comes from where $M_H = 1$ or some smaller value etc. leads to the conclusion that there is a large *inherent* ambiguity in such a bound and it is not clear how to assign a reliability estimate to such bounds.

In conclusion, we remind the reader of the well known fact that any prediction about the continuum theory made from measurements on a lattice from where the correlation length is finite (let alone of order unit lattice spacing) is inherently suspect. One needs much more analysis, perhaps on theories which are more rotationally invariant near $M_H = 1$, to pin down the error in M^H_{phys}.

Acknowledgements

This research was supported in part by the Florida State University Supercomputer Computations Research Institute which is partially funded by the Department of Energy through contract number DE-FC05-85ER250000. Without generous time on the Florida State University's new ETA[10] Supercomputer, this work would not have been possible. We wish to thank Herbert Neuberger for continuous encouragement and Peter Weisz for very useful discussions.

REFERENCES

1. R. Dashen and H. Neuberger, Phys. Rev. Letts., 50, (1983) 1897.

2. C. Whitmer, Princeton University Ph. D. Thesis, 1983, (unpublished); M. M. Tsypin, 'The Effective Potential of Lattice ϕ^4 Theory and the Upper Bound on the Higgs Mass', Lebedev Physical Institute preprint 280,1985; see also M. M. Tsypin in: Proceedings of the 'Lattice Higgs Workshop', Tallahassee FL, 16-18 May, 1988 (pub. World Scientific, 1988).

3. I. Montvay and W. Langguth, Z. Phys., C36, (1987) 725.

   A. Hasenfratz, T. Neuhaus, K. Jansen, Y. Yoneyama and C. Lang, Phys. Letts., 199B, (1987) 531.

4. K. Huang, E. Manousakis and J. Polonyi, Phys. Rev., D35, (1987) 3187.

   J. Kuti and Y. Shen, Phys. Rev. Letts., 60, (1988) 85.

   J Kuti, L. Lin and Y. Shen, Upper Bound on the Higgs Mass in the Standard Model, UCSD preprint UCSD/PTH 87–18.

   ibid., Fate of the Standard Model with a Heavy Higgs Particle, UCSD/PTH 87–19.

32

5. H. Neuberger, Soft Pions in Large Boxes, Rutgers University Preprint, RU-32-87.

   ibid, A Better Way to Measure f_π in the Linear Sigma Model, Rutgers University preprint, RU-42-87.

   U. M. Heller and H. Neuberger, The Finite Size Effects of Goldstone Bosons in Monte-Carlo Simulations, Santa Barbara preprint NSF-ITP-88-13; Rutgers University preprint RU-7-88.

   H. Neuberger, Phys. Letts., 199B, (1987) 536.

6. M. Luscher and P. Weisz, Nuclear Physics, B290 [FS20] (1987) 25, and Nuclear Physics B295[FS21] (1988)65; also see M. Luscher, 'Solution of the Lattice ϕ^4 Theory in 4 Dimensions', DESY 87-159, for a good description of the current philosophy and references.

7. P. Hasenfratz and J. Nager, Z. Phys., C37, (1988) 477.

8. R. Fukuda and E. Kyriakopoulos, Nucl. Phys., B85, (1975) 354.

   L O'Raifeartaigh, A. Wipf and H. Yoneyama, Nucl. Phys., B271, (1986) 653.

   M. Creutz and B Freedman, Annals of Physics, 132, (1981) 427.

TRIVIAL PURSUIT OF THE HIGGS PARTICLE

David J. E. Callaway

Department of Physics
The Rockefeller University
1230 York Avenue
New York, New York 10021-6399

ABSTRACT

The highly accurate Kadanoff lower-bound renormalization group for spin systems is generalized to models with local gauge symmetry. As an example it is applied to the Z_2 gauge-Higgs theory. The two critical exponents of the model are respectively predicted exactly and with 0.1% accuracy by a simple analytic calculation. The application of the technology to more complicated gauge groups in arbitrary spacetime dimension is described.

It is becoming evident that triviality plays an essential role in models of the weak interaction (see [1] and these proceedings). In particular, if a continuum limit of a Higgs lattice gauge theory exists, it may even be possible to predict the Higgs mass.[2,3]

One simple argument which suggests that the Higgs mass is predictable[1,2,3] is based upon the precepts of the renormalization group. If a continuum limit of a Higgs lattice gauge theory exists, it must correspond to a fixed point of the renormalization group transformation. The number of relevant directions at this fixed point equals the number of independent renormalized parameters in the theory. Yet universality suggests that the quartic coupling is irrelevant at this fixed point. This at least one parameter of the theory (e.g., the Higgs mass) might be predictable from the others.

Partially in order to explore this possibility, a Monte Carlo renormalization group analysis of the fixed-length $SU(2) \times U(1)$ standard model was performed.[3] This calculation was made using specific "maximal truncation" approximation.

Several fixed points do, in fact, appear in this scheme. Some of these fixed points possess "marginal" directions, suggesting that (in addition to the possibility of a calculable quartic coupling, as implied by universality) *a priori* bounds on other renormalized couplings might be set. Before a definitive staement can be made, however, an accurate real-space renormalization group technique for lattice Higgs systems is needed. One candidate technique,[4] a generalized lower-bound renormalization group, is discussed here.

The Kadanoff lower-bound renormalization group (LBRG) predicts[5] critical exponents of spin systems with a precision not achievable by any other simple RG method (see [6] for a review). The LBRG gives $dv = 2 - a$ to 0.1% accuracy in the $d = 2$ and 3 Ising models, and to 1% accuracy in four dimensions. For ϕ^4 field theory it yields a one-dimensional integral equation whose solution give ν correctly to first order in the ε-expansion. It is also a direct approximation to an infinite system, so true finite-size effects are absent.

This amazing precision, coupled with the present level of interest in the renormalization-group structure of gauge-Higgs systems (see, e.g., [1]), provides motivation to generalize this simple technique to systems with local gauge symmetry. The generalization presented here is applicable in arbitrary spacetime dimension, and reduces to the Kadanoff method for spin systems in the limit where the gauge coupling vanishes.

The crux of the LBRG is the observation that [since $\langle \exp(\Delta S) \rangle \geq \exp(\langle \Delta S \rangle)$] the addition to the action of an operator with vanishing expectation value *lowers* the free energy f. Interaction-moving operations satisfy this criterion by translational invariance. Variational parameters are introduced and are (easily) optimized at a fixed point to give a best lower bound for f. Critical exponents then follow directly from the recursion relations.

The application of the original LBRG to gauge theories is complicated by the fact that its block lattice points lie in the centers of hypercubes, while the gauge links $\{U\}$ are only defined along its edges. It is therefore first necessary to use a prefacing transformation to map the original "plaquette" system to a "subsumed" model which allows parallel transport to the center of a hypercube. The blocking then yields a plaquette model with larger latic spacing, ready for the next iteration.

An example facilitates explanation. Consider a single plaquette in a two-

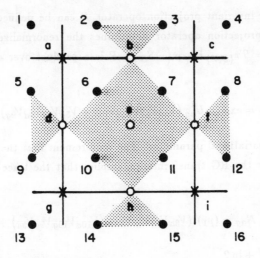

Figure 1. Definition of notation

dimensional Z_2 gauge theory with vertices labelled (1,2,3,4) and action $S_{G,1234} = -\beta U_{12}U_{23}U_{34}U_{14}$ where $U_{ji} = U_{ij} = \pm 1$. Place a point c in the center of the plaquette. The corresponding subsumed model action is

$$\tilde{S}_{G,1234} = -\tilde{\beta}\left(V_{1c}U_{12}V_{2c} + V_{2c}U_{23}V_{3c} + V_{3c}U_{34}V_{4c} + V_{1c}U_{14}V_{4c}\right) - C_\beta \quad (1)$$

where each of the $v_{ic} = \pm 1$ is an element of Z_2 which runs between points c and $i = 1$ to 4. Since $Tr_{\{V\}}\left[\exp\left(-\tilde{S}\right)\right] = \exp(-S)$ it follows that $\left(\tilde{\beta}\right)'' = \beta$ and $C_\beta = -2\left(\tilde{\beta}\right)' - \beta - \ln 16$, where $x'' \equiv (x')'$ and $\tanh x' \equiv \tanh^2 x$. In higher spacetime dimensions, the point c is placed in the center of the hypercube. The prefacing transformation can be performed analytically for discrete groups. For larger groups more interactions [e.g., $(VUV)^2, (VUV)(VUV)$ etc.] must be included. Note that the prefacing is *underconstrained*—many subsumed models correspond to one plaquette model.

The machinery of the generalized LBRG can be illustrated using a Z_2 gauge theory in $d = 2$ dimensions. Figure 1 defines the notation. The original lattice fields (e.g., U_{67}) run between the original site points, denoted by numerals 1–16. The new fields of the subsumed model (e.g., V_{6e}) go between the original site points and new points labelled by letters a through i. Block points are denoted by crosses, and block fields (e.g., U'_{ag}) connect these. Thus (a, c, i, g) is the boundary of a typical block of side length $b = 2$, which is the RG scaling factor.

A gauge-invariant projection operator P can be defined for the subsumed model. The projection operator determines the renormalized action $S_{G,R}$ via $\exp\left(-S_{G,R}\right) = Tr_{\{U,V\}} P \exp\left(-\tilde{S}_G\right)$. P is a product over all block links $\{U'\}$ of terms like

$$P_{ag} = \exp\left[p_1 U'_{ag}\left(V_{6a}V_{6d}V_{10d}V_{10g} + V_{5a}V_{5d}V_{9d}V_{9g}\right) - N_{ag}\right] \tag{2}$$

where p_1 is a variational parameter. The requirement that the free energy remain invariant under the RG transformation implies that the trace of P_{ag} over U'_{ag} is unity, and thus

$$N_{ag} = (p_1)'\left(V_{6a}V_{6d}V_{5d}V_{5a}\right)\left(V_{10d}V_{10g}V_{9g}V_{9d}\right) + C_1$$

with $C_1 \equiv (p_1)' + \ln 2$.

So far the RG transformation is *exact* (and intractable). The LB approximation is to move all interactions from the VUV terms and from the normalization terms (e.g., N_{ag}) in the product $P \exp(-\tilde{S}_G)$ into the shaded region (b, f, h, d), and equally to its counterparts in other blocks. The result for the contribution from block (a, c, i, g) is

$$
\begin{aligned}
\left[P\exp\left(-\tilde{S}_G\right)\right]_{LB,acig} = \exp\Big\{ &2\tilde{\beta}\left[U_{6,7}\left(V_{6b}V_{7b} + V_{6e}V_{7e}\right)\right. \\
&+ U_{7,11}\left(V_{7f}V_{11f} + V_{7e}V_{11e}\right) \\
&+ U_{10,11}\left(V_{10e}V_{11e} + V_{10h}V_{11h}\right) \\
&+ \left.U_{6,10}\left(V_{6e}V_{10e} + V_{6d}V_{10d}\right)\right] - 4C_\beta \\
&+ p_1\Big[U'_{ag}\left(V_{6a}V_{6d}V_{10d}V_{10g}\right) \\
&+ U'_{gi}\left(V_{10g}V_{10h}V_{11h}V_{11i}\right) \\
&+ U'_{ci}\left(V_{7c}V_{7f}V_{11f}V_{11i}\right) + U'_{ac}\left(V_{6a}V_{6b}V_{7b}V_{7c}\right)\Big] \\
&- p'_1\left[\left(V_{6b}V_{7b}V_{6e}V_{7e}\right)\left(V_{10e}V_{11e}V_{10h}V_{11h}\right)\right. \\
&+ \left.\left(V_{6d}V_{6e}V_{10d}V_{10e}\right)\left(V_{7e}V_{7f}V_{11e}V_{11f}\right)\right] \\
&- 2C_1\Big\}
\end{aligned}
\tag{3}
$$

The interaction-moving separates the original system into blocks within which the summations over $\{U, V\}$ can be performed independently. The renormalized couplings β_R and C_R are found from

$$\exp\left[\beta_R\left(U'_{ag}U'_{gi}U'_{ci}U'_{ac}\right) + C_R\right] = \left[\exp\left(-S_{G,R}\right)\right]_{acig}$$
$$= Tr_{\{U,V\}}\left[P\,\exp\left(-\tilde{S}_G\right)\right]_{LB,acig} \tag{4}$$

The variational parameter p_1 is determined from the extremum condition

$$0 = \frac{\partial f}{\partial p_1} = \frac{\partial f}{\partial K_\alpha}\frac{\partial K_\alpha}{\partial p_i} \tag{5}$$

where $\alpha = 1,\ldots,n; K_{R,n} \equiv C_R$. This can be solved easily at a fixed point,[1] for there $\partial f/\partial K_\alpha \equiv e_\alpha$ is a left eigenvector of the matrix $D_{\alpha\beta} \equiv \partial K_{R,\alpha}/\partial K_\beta$ with eigenvalue $b^d \left(= 2^2\right)$. In fact, Eq. (5) is then a determinant, since e_α is proportional to $\mathrm{cof}\left(\Delta_{an}\right), \Delta_{\beta\alpha} \equiv D_{\alpha\beta} - b^d\delta_{\alpha\beta}$. The result at the infinite β fixed point is $\beta_R = \left(\frac{1}{2}\left(2\tilde{\beta}\right)''\right)''$, $(p_1) = \frac{1}{2}\left(2\tilde{\beta}\right)$ ", giving the exact limit $\partial\beta_R/\partial\beta \to 1 = b^y$. Thus $y = 0$ (the fixed point is *marginal*). No fixed points exist at finite β. Exact blocking yields $\beta_R = \left(\left(\tilde{\beta}\right)''\right)'' = \beta''$ by comparison.

Scalar "Higgs" fields are easily included in the formalism. Consider for simplicity fixed-length Z_2 fields $\sigma_n = \pm 1$ (variable-length scalars are treated in [1]). The Higgs terms in the original action S as well as in the renormalized action S_R can be written entirely in terms of gauge-invariant objects like $h_{6,7} \equiv \sigma_6 U_{6,7}\sigma_7$ and $h'_{ac} \equiv \sigma'_a U'_{ac}\sigma'_C$ respectively. For scaling factor $b = 2$ in two dimensions nothing larger than a plaquette can be included. Thus the allowable Higgs terms for plaquette $(6,7,10,11)$ are:

$$O_{Plaq} \equiv h_{6,7}\,h_{7,11}\,h_{10,11}\,h_{6,10}$$
$$O_2 \equiv \frac{1}{2}\left(h_{6,7} + h_{7,11} + h_{10,11} + h_{6,10}\right)$$
$$O_3 \equiv \frac{1}{2}h_{6,7}\left(h_{7,11} + h_{6,10}\right)\left(1 + O_{Plaq}\right) \tag{6}$$
$$O_4 \equiv 2\,O_{Plaq}\,O_2$$
$$O_5 \equiv \frac{1}{2}h_{6,7}\,h_{10,11}\left(1 + O_{Plaq}\right)$$

and their contribution to the action is $-\Sigma_{i=2}^5 K_i O_i$. The Higgs contribution to the full action is the sum over all plaquettes of this quantity (note that $\beta \equiv K_1$ and likewise $\beta_R \equiv K_{R,1}; C_R \equiv K_{R,6}$ etc.). The projection operator Q for the scalars is a product over all block fields of terms like this for σ_a':

$$Q_a = \exp\left[p_2\sigma_a' \sum - \frac{1}{2}L_2\left(\sum^2 - 4\right)\right.$$
$$\left. - M_2\,\sigma_1\,\sigma_2\,\sigma_5\,\sigma_6\,V_{1a}\,V_{2a}\,V_{5a}\,V_{6a} - C_2\right]$$
$$\sum \equiv V_{1a}\,\sigma_1 + V_{2a}\,\sigma_2 + V_{5a}\,\sigma_5 + V_{6a}\,\sigma_6$$

(7)

where the normalization $Tr_{\sigma'}Q = 1$ implies that

$$4L_2 = (2p_2)',\ 2M_2 = (2L_2)',\ -C_2 = -2L_2 + M_2 - \ln 2.$$

The LB approximation of moving all interactions equally into the region (b, f, h, d) and its counterparts in other blocks is made as before. The LB contribution to $\exp(-S_R)$ from block (a, c, i, g) is the sum over all contained $\{U, V\}$ of

$$\left[PQ\,\exp\left(-\tilde{S}\right)\right]_{LB,acig} = \exp\left\{4\left(2K_2O_2 + K_3O_3 + K_4O_4 + K_5O_5\right)\right.$$
$$- \frac{1}{2}L_2\left[(\sigma_6\,V_{6e} + \sigma_7\,V_{7e} + \sigma_{10}\,V_{10e} + \sigma_{11}\,V_{11e})^2 - 4\right]$$
$$- M_2\,(\sigma_6\,\sigma_7\,\sigma_{10}\,\sigma_{11}\,V_{6e}\,V_{7e}\,V_{10e}\,V_{11e}) - C_2$$
$$\left. + p_2\left(\sigma_a'\,V_{6a}\,\sigma_6 + \sigma_c'\,V_{7c}\,\sigma_7 + \sigma_i\,V_{11i}\,\sigma_{11} + \sigma_g'\,V_{10g}\,\sigma_{10}\right)\right\}$$
$$\times\left[P\,\exp\left(-\tilde{S}_G\right)\right]_{LB,acig}$$

(8)

The six block couplings $\{K_R\}$ are evaluated analytically in terms of the $\{K\}$, and the p_i are determined by Eq. (5). The familiar[5] Ising fixed point is recovered at $p_2 = 0.76$ and infinite β and p_1, yielding $d\nu = 2 - \alpha = 1.998$ to 0.1% accuracy. No distinct new fixed points appear at finite β [β always decreases when Eqs. (5) are applied].

Thus it is seen that by the use of a prefacing transformation the accurate Kadanoff LBRG can in fact be applied to systems with local gauge symmetry. In the case of the two-dimensional Z_2-Higgs theory, the two critical exponents are respectively determined exactly and to 0.1% accuracy by a simple analytical calculation. The method is applicable to more complicated gauge groups in higher spacetime dimension, though numerical techniques[7] may then be needed. Often however an invariant subspace exists which can vastly simplify the calculation (for instance, the space of plaquette products on opposite sides of the cube for a Z_2 gauge theory in three dimensions). Moreover in all cases the generalization does reduce to the original technology[5] for spin systems in the limit of vanishing gauge coupling. Thus the good results for spin systems are automatically recovered as a special case.

Acknowledgements

It is a pleasure to thank R. Petronzio for informative discussions. This work was supported in part by the U.S. Department of Energy contract DE–AC02–87–ER40325 – Task B_1.

REFERENCES

1. Callaway, D. J. E., Triviality Pursuit: Can elementary scalar particles exist?, Rockefeller preprint RU/87/B_1/20, to be published in Physics Reports.

2. Callaway, D. J. E., Nucl. Phys., B233, (1984) 189.

3. Callaway, D. J. E. and R. Petronzio, Nucl. Phys., B292, (1987) 497.
   Callaway, D. J. E. and R. Petronzio, Nucl. Phys., B240 [FS12], (1984) 577.
   Callaway, D. J. E. and R. Petronzio, Nucl. Phys., B267, (1986) 253.

4. Callaway, D. J. E., Rockefeller preprint RU/87/B_1/27.

5. Kadanoff, L. P., Phys. Rev. Lett., 34, (1975) 1005.
   Kadanoff, L. P., A. Houghton, M. C. Yalabik, J. Stat. Phys., 14, (1976) 171.

6. Burkhardt, T. W.,Real-Space Renormalization, eds. T. W. Burkhardt and J. M. J. van Leeuwen (Springer-Verlag, New York, 1982).

7. Callaway, D. J. E., and R. Petronzio, Phys. Lett., 139B, (1984) 189.
   Callaway, D. J. E., and R. Petronzio, Phys. Lett., 145B, (1984) 381.
   Callaway, D. J. E., and R. Petronzio, Phys. Lett., 148B, (1984) 445.
   Callaway, D. J. E., and R. Petronzio, Phys. Lett., 149B, (1984) 175.

LOW ENERGY THEOREMS, UNITARITY, AND EXPERIMENTAL PROSPECTS FOR THE SYMMETRY BREAKING SECTOR

Michael S. Chanowitz

Lawrence Berkeley Laboratory
University of California
Berkeley, California 94720, U.S.A.

ABSTRACT

Low energy theorems are derived for scattering of longitudinally polarized W and Z's, providing the basis for an estimate of the observable signal if electroweak symmetry breaking is due to new physics at the TeV scale. A pp collider with $\sqrt{s}, \mathcal{L} = 40$ TeV, $10^{33} cm^{-2} s^{-1}$ is just sufficient to observe the signal while pp colliders with 40, 10^{32} or 20, 10^{33} are not. A collider that is sensitive to the TeV-scale signal provides valuable information about symmetry breaking whether the masses of the associated new particles are below, within, or above the 1–2 TeV region.

1. INTRODUCTION

I am pleased to be able to participate in this workshop. The lattice is clearly a very useful tool for the study of strongly coupled symmetry breaking sectors for the electroweak theory. The study of the triviality of the minimal Higgs sector is an interesting subject in itself and useful as a "warm-up exercise". There are many other strong-coupling models that could benefit from lattice studies (see reference 1 for a review with additional references). The first example that comes to mind is technicolor (reviewed by Eichten at this meeting), where lattice studies could check the present crude estimates based on large N_{Color} scaling laws. The slowly running technicolor variants[2] could also benefit from examination by lattice methods. In the scalar sector of the standard model one could investigate the hypothesis of dynamical generation of vector bosons, that is inspired by two dimensional CP(N) models but is only conjectured in four dimensional models.[3] The basic idea of ultracolor[4] could also be tested—that it is possible to obtain the minimal Higgs sector as a low energy approximation to a confining gauge theory that can be chosen (so far only unnaturally) at a higher mass scale than the usual technicolor scale. Finally, composite technicolor,[5] and composite models in general, are clearly rich territory for lattice studies. Of course, except for the question of dynamical vector boson generation in the scalar sector, the problems of fermions on the lattice will play an important role in these studies.

When there is a sequel to this meeting it would be useful to hear from people working on these models. I will not discuss them but will instead present a model-independent analysis that goes back to the 60's, when we were looking for a strong interaction theory that we couldn't have solved if we'd found it. The approach was based on symmetry properties, as reviewed in the *Current Algebras* book by Adler and Dasher.[6] The other ingredient of the analysis is unitarity, going back even further to the studies[7] of Lee and Yang and of Ioffe, Okun, and Rudik, who obtained bounds on the scale of the weak interaction quanta from the Fermi theory. Here we use general symmetry and unitarity properties to bound the scale of the quanta in the symmetry breaking sector of the electroweak theory.

Within the general framework of spontaneous symmetry breaking—the only known way of constructing a sensible broken gauge theory—we believe that the W and Z masses must result from new, unknown particles that interact by a new unknown force. We know neither the mass scale M_{SB} of the new particles ("SB" for symmetry breaking) nor the strength λ_{SB} of the new force. With the SSC we will be able to determine whether the new force is a strong interaction

$$\frac{\lambda_{SB}}{8\pi} \cong 0(1) \tag{1}$$

or a weak interaction

$$\frac{\lambda_{SB}}{8\pi} \ll 1. \tag{2}$$

In the first case I will argue that the new physics lies above 1 TeV. In this case it is also likely that the new spectrum begins within or near the 1–2 TeV region and will be directly observable. In the second, weak coupling case the new particles are much lighter than the 1 TeV scale, *i.e.*, a few hundred GeV or below, and will be copiously produced at the SSC. These statements taken together are what I call the "No-Lose Corollary".

The basic physical point is that the longitudinal polarization modes of the W and Z, denoted W_L and Z_L, are actually degrees of freedom that originate, by the Higgs mechanism, in the symmetry breaking sector. They are essentially the "pions" of the symmetry breaking sector, and like the pions of hadron physics they obey low energy theorems characteristic of the scattering of Goldstone bosons.[10] If there are no other light (compared to 1 TeV) particles in the symmetry breaking sector, the low energy scattering amplitudes depend only on the known parameters G_F and $\rho = (M_W/M_Z \cos\theta_W)^2$ and not at all on the unknown physics of the symmetry breaking sector, denoted by its Lagrangian $\mathcal{L}_{SB}$.[11] For example, one of the low energy amplitudes is

$$\mathcal{M}(W_L^+ W_L^- \to Z_L Z_L) = \sqrt{2} G_F \frac{s}{\rho}. \tag{3}$$

The low energy theorem provides the correlation between the mass scale M_{SB} and the interaction strength λ_{SB}. Unitarity requires that the amplitude cannot be proportional to s for arbitrarily large s, and the most likely scenario, discussed below, is that the growth in s is cut off at the mass scale M_{SB}. This observation allows us to correlate the strong coupling regime, eq. (1), with the mass domain $M_{SB} \gtrsim 1$ TeV (c.f. section 3).

For the SSC the crucial observation is that strong W_L, Z_L scattering is observable at the SSC by virtue of increased yields of gauge boson pairs produced with the WW fusion mechanism.[12] At the design energy and luminosity these extra gauge boson pairs will be observable above the background sources of gauge boson pairs that are present whether $\mathcal{L}_{SB}$ is a strong or weak coupling theory. The conclusion is the

No-Lose Corollary:

> *Either there are light ($\ll$ 1 TeV) particles from $\mathcal{L}_{SB}$ that can be produced and studied directly*

and/or

> *Excess WW, WZ, ZZ production is observable, signaling strongly-coupled $\mathcal{L}_{SB}$ with $M_{SB} \gtrsim 1$ TeV.*

For the strong coupling case, if as in hadron physics resonances occur when the partial wave amplitudes are $O(1)$, then probably $M_{SB} \lesssim 0(2)$ TeV and the low-lying spectrum of $\mathcal{L}_{SB}$ is (just) visible. However, the 1–2 TeV strong coupling signal would be observable even if $M_{SB} >> 2$ TeV and the new particles were too heavy to produce.

In the weak coupling case there is one known exception that should be mentioned: if $\mathcal{L}_{SB}$ is given by the minimal Higgs model and if the mass happens to lie in the interval $2m_t < m_H < 2M_W$, then although 10^6 Higgs would be produced in an SSC year, it is not now known how to detect them in their $H \to \bar{t}t$ decay mode.[13] For now the only known way of discovering the Higgs boson in this mass range is to build a $\sqrt{s} = 300$ GeV e^+e^- collider with a luminosity of $\sim 10^{32}$ cm^{-2} s^{-1}. Even in this rather unlikely scenario, the SSC would contribute to our understanding of symmetry breaking by verifying the absence of the excess gauge boson pairs associated with new strong interactions. (Strong coupling models might have light scalars that would approximately mimic a light Higgs boson.) In general the absence of additional gauge boson pairs from WW fusion would be our cue for a redoubled search of the sub-TeV mass scale.

I want to make a few comments to put the above statements in perspective. TeV scale symmetry breaking physics puts the greatest demands on energy and luminosity. Most other physics that has been contemplated is less demanding and less sensitive to the difference between 20 and 40 TeV. However, symmetry breaking physics has a special claim on our attention, since while other possible new physics (e.g., SUSY, Z', $\cdots$) may or may not occur in nature, we know for sure that the electroweak symmetry is broken. The no–lose corollary says that a facility able to "see" the 1–2 TeV signal of strongly–coupled symmetry breaking gives us valuable information about the mass scale M_{SB} of the symmetry breaking physics whether the new particles occur below, within, or above the 1–2 TeV region. As will become clear from the results presented below, a pp collider with $\sqrt{s} = 40$ TeV and $\mathcal{L} = 10^{33} cm^{-2} s^{-1}$ is *just* able to observe the 1–2 TeV strong–interaction signal but is not sufficient to study it in detail. In this sense it is an optimal probe, since we would not want to consider a more ambitious facility without first knowing that there is TeV–scale physics to study.

Though the problem of observing the strong interaction symmetry breaking signal at TeV e^+e^- colliders has not been as extensively studied, several authors have had a first look at the signal and/or backgrounds.[14] The conclusion from Bento and Llewellyn–Smith is that the signal of the conservative model discussed below would not be visible at an e^+e^- collider with $\sqrt{s}, \mathcal{L} = 2$ TeV, $10^{33} cm^{-2} s^{-1}$ but requires a collider in the range $\sqrt{s}, \mathcal{L} \sim 3 - 5$ TeV, $(1 - 2\frac{1}{2})10^{33} cm^2 s^{-1}$ Incidentally, though we are accustomed to e^+e^- colliders being cleaner than hadron colliders, it is amusing in this instance that the dominant background to WW fusion at an e^+e^- collider, from the usual two photon process, $\gamma\gamma \to WW$, is relatively unimportant at pp colliders (because of the quark and electron charges that contribute to the fourth power and also because the smaller electron mass enhances the factor $[\ln E/m_e]^4$). On the other hand the dominant background at the pp colliders, from $\bar{q}q \to WW$ as discussed below, has no counterpart in e^+e^- collisions.

The remainder of the talk is organized as follows: Section 2 reviews relevant aspects of spontaneous symmetry breaking and sketches a proof of the low energy theorems using current algebra methods borrowed from hadron physics.[11] Section 3 is a discussion of unitarity and a conservative strong interaction model that is inspired by the low energy theorems.[15] In Section 4 I will discuss the experimental signals for a strongly interacting symmetry breaking sector that could be seen at the SSC. Work in this area continues and is far from complete. A brief conclusion is presented in Section 5.

2. SPONTANEOUS SYMMETRY BREAKING AND LOW ENERGY THEOREMS

In order to implement spontaneous symmetry breaking, the Lagrangian of the symmetry breaking sector, $\mathcal{L}_{SB}$, must possess a global symmetry group G—analogous to the flavor symmetry of QCD—which breaks by asymmetry of the vacuum to a smaller group H,

$$G \to H. \tag{4}$$

Gauge invariance requires that G include the electroweak $SU(2)_L \times U(1)_Y$ and that H include the unbroken electromagnetic U(1). For each broken generator of G there is a massless goldstone boson in the spectrum of $\mathcal{L}_{SB}$. Three of these couple to the weak currents and are denoted $w^\pm$, z. Others, if any, are denoted by $\{\phi_i\}$. Including the electroweak gauge interactions, the goldstone triplet $w^{\pm}$, z become longitudinal gauge boson modes $W_L^{\pm}$, Z_L, and the $\{\phi_i\}$ acquire small masses $0(gM_{SB})$, becoming "pseudo-goldstone" bosons.

As an example, for two flavor QCD with massless quarks the global symmetry G is $SU(2)_L \times SU(2)_R$. After spontaneous symmetry breaking the surviving invariance group is $H = SU(2)_{L+R}$ which is just the isospin group. There are three broken generators, corresponding to the axial generators $SU(2)_{L-R}$, so that three massless goldstone bosons emerge, $\pi^{\pm}$ and π^0. If there were no other symmetry breaking physics, $\mathcal{L}_{SB}$, $\pi^{\pm}$ and π^0 would indeed become longitudinal modes of $W^{\pm}$ and Z, which would however have masses of order 40 MeV rather than $\sim$ 100 GeV. (This would have made a rather different world than the one we live in!)

The statement that the longitudinal modes $W_L^{\pm}$, Z_L are identified with the goldstone bosons $w^{\pm}$, z is given a precise meaning by the equivalence theorem, proved to all orders in ref. [15]:

$$M\left(W_L(p_1)W_L(p_2)\ldots\right)_U = M\left(w(p_1)w(p_2)\ldots\right)_R + O\left(\frac{M_W}{E_i}\right). \qquad (5)$$

In eq. (5) the left side is an S-matrix element involving longitudinal modes in the U or unitary gauge while the right side is the corresponding goldstone boson amplitude in an R or renormalizeable gauge. As indicated in eq. (5), the equivalence holds at energies large compared to the W and Z masses. We can use the equivalence theorem to translate statements about goldstone boson scattering amplitudes into statements about scattering of longitudinally polarized W's and Z's.

As an immediate application, consider the case[15] in which the global symmetry G includes $SU(2)_L \times SU(2)_R$ and H includes an $SU(2)_{L+R}$. For such theories $\rho = 1$ up to electroweak corrections and we may immediately apply the pion low energy theorems[10] derived from current algebra for just this case. For instance, just as for pions we have

$$M(\pi^+\pi^- \to \pi^0\pi^0) = \frac{s}{f_\pi^2} \qquad s \ll 1 \text{ GeV}^2 \qquad (6)$$

for $\mathcal{L}_{SB}$ with no other particles that are light compared to M_{SB} we would have

$$M(w^+w^- \to zz) = \frac{s}{v^2} \qquad s \ll M_{SB}^2 \qquad (7)$$

where $v = 0.25$ TeV is the familiar vacuum expectation value, $M_W = \frac{1}{2}gv$. With the equivalence theorem this becomes a statement about the scattering of W_L and Z_L in an intermediate energy domain:

$$M(W_L^+ W_L^- \to Z_L Z_L) = \frac{s}{v^2} \qquad M_W^2 \ll s \ll M_{SB}^2. \tag{8}$$

Eq. (8) and eq. (3) are equal up to small corrections, $0(M_W^2/M_{SB}^2)$, for $\rho = 1$.

The assumptions used above, $G \supset SU(2)_L \times SU(2)_R$ and $H \supset SU(2)_{L+R}$, are sufficient to guarantee $\rho = 1$ to all orders in λ_{SB} but they are not known to be necessary conditions. We are therefore motivated to derive the low energy theorems for all candidate groups G and H and for all values of ρ. The problem we face is equivalent to that of obtaining the pion–pion scattering low energy theorems in the absence of isospin symmetry. We derive the low energy theorems by three different methods:[11] a perturbative power counting analysis, nonlinear chiral Lagrangians, and current algebra. I will sketch the current algebra derivation below. Along with the low energy theorems for general values of ρ, the derivation establishes a kind of converse to the result quoted above: we find that if $\rho = 1$ then the goldstone boson sector consisting of $w^\pm$, z possesses an effective $SU(2)_{L+R}$ symmetry ("custodial" SU(2)) in the low energy domain $s \ll M_{SB}^2$.

Briefly the derivation is as follows. The global symmetry G must be at least as large as the gauge group, $G \supset SU(2)_L \times U(1)_Y$, so in particular we have the $SU(2)_L$ charge algebra

$$[L_a, L_b] = i\varepsilon_{abc}L_c \tag{9}$$

where the corresponding local currents L_a^μ can generally be expanded in terms of the goldstone triplet $w^\pm$, z as

$$L_a^\mu = \frac{1}{2}r_a\varepsilon_{abc}w_b\partial^\mu w_c - \frac{1}{2}f_a\partial^\mu w_a + \dots \tag{10}$$

with terms involving heavy fields omitted. Since $H \supset U(1)_{EM}$ we have $f_1 = f_2$ and $r_1 = r_2$. The f_a are analogues of the PCAC constant and determine the gauge boson masses,

$$M_W = \frac{1}{2}gf_1, \tag{11}$$

$$\rho = (f_1/f_3)^2. \tag{12}$$

Corrections are suppressed by inverse powers of order M_{SB} or, because of quantum corrections, by inverse powers of $4\pi f_a$.

It is straightforward to show that the $SU(2)_L$ algebra requires

$$r_1 = \frac{1}{\sqrt{\rho}}, \tag{13}$$

$$r_3 = 2 - \frac{1}{\rho}, \tag{14}$$

so that the parameters r_a and f_a in eq. (10) are completely determined in terms of G_F and ρ. In particular, $\rho = 1$ implies $f_1 = f_2 = f_3$ and $r_1 = r_2 = r_3 = 1$ which means that the goldstone boson contributions to L_a^μ are the difference of $SU(2)$ vector and axial vector currents. The existence of this vector $SU(2)$ triplet of currents establishes the converse alluded to above.

The rest of the derivation is much like the usual current algebra derivation[10] except that we do not assume an $SU(2)_{L+R}$ isospin invariance. Consequently pole terms which are forbidden by G-parity in the pion case are not forbidden here. Assuming that $w^\pm$, z saturate these pole terms we find goldstone boson low energy theorems such as

$$M(w^+ w^- \to zz) = \frac{s}{f_1^2} \frac{1}{\rho} \qquad s \ll M_{SB}^2 \tag{15}$$

which, using (11), reduces to (8) for the case $\rho = 1$. By the equivalence theorem we have then

$$M(W_L^+ W_L^- \to Z_L Z_L) = \frac{s}{v^2} \frac{1}{\rho} \qquad M_W^2 \ll s \ll M_{SB}^2 \tag{16}$$

with $v = f_1 \cong 2M_W/g$. The other two independent amplitudes are

$$M(W_L^+ W_L^- \to W_L^+ W_L^-) = -\frac{u}{v^2} \left(4 - \frac{3}{\rho} \right), \tag{17}$$

$$M(Z_L Z_L \to Z_L Z_L) = 0, \tag{18}$$

and by crossing we have also

$$M(W_L^\pm Z_L \to W_L^\pm Z_L) = \frac{t}{v^2}\frac{1}{\rho}, \tag{19}$$

$$M(W_L^+ W_L^+ \to W_L^+ W_L^+) = M(W_L^- W_L^- \to W_L^- W_L^-) = -\frac{s}{v^2}\left(4 - \frac{3}{\rho}\right). \tag{20}$$

Like (16), eqs. (17–20) are valid in the intermediate domain $M_W^2 \ll E_i^2 \ll M_{SB}^2, (4\pi v)^2$.

3. UNITARITY AND A CONSERVATIVE MODEL

It would be difficult to test the low energy theorems directly at a pp collider because of the predominance of the $\bar{q}q \to WW$ background for $s \ll 1$ TeV^2. The only hope is to use the polarization information since the signal consists predominantly of longitudinally polarized gauge boson pairs while the background is predominantly transversely polarized. This seems unlikely as a first generation experiment, though it has not yet been examined carefully.

In any case, at present a more interesting application is to use the low energy theorems to estimate the generic signal we should expect if $\mathcal{L}_{SB}$ is a strongly coupled sector at the TeV scale or above. The problem we face is like the one that physicists of the 1930's would have faced if they knew nothing of nuclei, baryons or other hadrons, but had discovered the pion, measured the PCAC constant f_π, and recognized (!?) the pion as an almost Goldstone boson. They would have then been able to derive the pion–pion low energy theorems, such as eq. (6), and the problem would be to use this information to reconstruct the scale of hadron physics. Though it would take a skilled writer of science fiction to make this a plausible plot line for the 1930's, it is precisely the situation we are in today if $\mathcal{L}_{SB}$ is strongly coupled: our pions are the longitudinal modes of W and Z, our "PCAC" constant is $v = 0.25$ TeV, and we have the low energy theorems, eqs. (16–20).

The central ingredient in our considerations is unitarity. The linear growth in s, t, u of the amplitudes (16–20) cannot continue indefinitely or unitarity would be violated. For instance the $W_L W_L \to Z_L Z_L$ amplitude (16) is pure s-wave. If we adopt the low energy amplitude (16) as a model of the absolute value of the scattering amplitude, then the $J = 0$ partial wave amplitude is

$$\left|a_0(W_L^+ W_L^- \to Z_L Z_L)\right| = \frac{s}{16\pi v^2} \tag{21}$$

where here and hereafter we set $\rho = 1$. Unitarity requires $|a_0| \leq 1$ so we see that the growth of a_0 must be cut off at a scale Λ with

$$\Lambda \leq 4\sqrt{\pi}v = 1.8 \text{ TeV}. \tag{22}$$

At the cutoff $\sqrt{s} = \Lambda$ the order of magnitude of the amplitude is

$$|a_0(\Lambda)| = \frac{\Lambda^2}{16\pi v^2}. \tag{23}$$

For $\Lambda \lesssim v \cong \frac{1}{4}$ TeV we have $|a_0(\Lambda)| \ll 1$ indicating a weakly interacting theory for the symmetry breaking dynamics $\mathcal{L}_{SB}$, while for $\Lambda \gtrsim 1$ TeV we have $|a_0(\Lambda)| \cong 0(1)$, the hallmark of a strong interaction theory. Though there is one counterexample mentioned below, the most likely dynamics is that the cutoff scale Λ is of the order of M_{SB}, the mass scale of the new quanta. Then for $\Lambda \cong 0(M_{SB})$ eq. (23) establishes the relationship mentioned in the introduction between the mass scale of the new quanta and the strength of the new interactions: weak coupling for $M_{SB} \ll 1$ TeV and strong coupling for $M_{SB} \gtrsim 0(1)$ TeV.

A weak coupling example is provided by the standard Higgs model[8] with a light Higgs boson, $m_H \ll 1$ TeV, which can be treated perturbatively. Then $a_0(W_L W_L \to Z_L Z_L)$ is given in tree approximation by (where I neglect M_W^2/s)

$$a_0 = \frac{-s}{16\pi v^2} \frac{m_H^2}{s - m_H^2}. \tag{24}$$

For $s \ll m_H^2$ this agrees with the low energy theorem (22) while for $s \gg m_H^2$ it saturates at the constant value $m_H^2/16\pi v^2$. Comparing with (23) we see that m_H indeed provides the scale for Λ.

A strong interaction example is provided by hadron physics. For the $J = I = 0$ partial wave, the low energy theorem[10] gives

$$a_{00}(\pi\pi \to \pi\pi) = \frac{s}{16\pi f_\pi^2} \tag{25}$$

with $f_\pi = 92$ MeV. Eq. (25) saturates unitarity at $4\sqrt{\pi}f_\pi = 650$ MeV which is indeed the order of the hadron mass scale. The a_{11} and a_{02} amplitudes saturate at 1100 and 1600 MeV.

For weak coupling the partial wave amplitudes saturate at values small compared to 1 giving rise to narrow resonances at masses well below 1 TeV. For strong coupling they saturate the unitarity limit with broad resonances in the TeV range.

There is one strong coupling model that lacks well-defined O(TeV) resonances, namely the 0(2N) Higgs model solved to leading order for $N \to \infty$ and then evaluated at $N = 2$ which corresponds to the standard model.[16] (This is only a little worse than the large N_{color} limit for QCD which approximates $3 \cong \infty$; I admit to uneasiness with both approximations.) In that model, which is a strong coupling model, the low energy theorems are of course valid and there is indeed a slow (logarithmic) saturation of partial wave unitarity at the TeV scale, but there are no discernible resonances in the TeV region.

The 0(2N) model, or the possibility (which cannot be definitively excluded by the heuristic estimates given above) that resonance structure might be deferred to 2 TeV or above, both motivate a conservative model for strong interactions that Mary Gaillard and I have considered.[15] In this model we represent the absolute values of the partial wave amplitudes by the low energy theorems up to the energy at which unitarity is saturated and set them equal to one for higher energies. The model is conservative in three respects:

- it neglects possible (or, I would say, likely) resonance structure, underestimating the yield for the 1 TeV Higgs boson by $\sim 50\%$.

- it neglects higher partial waves which surely begin to contribute as the lowest waves saturate.

- it correctly represents the order of magnitude seen in $\pi\pi$ data.

We discuss the experimental implications in the next section.

4. EXPERIMENTAL SIGNALS FOR STRONG INTERACTION MODELS

We consider what might actually be observed at the SSC if $\mathcal{L}_{SB}$ is a strongly interacting theory. The generic strong interaction signal is longitudinally polarized W, Z pairs produced by WW fusion. It will help to measure the polarization of the gauge bosons[17] (if statistics are sufficient) but I will *not* assume polarization information in the results given here. Then the irreducible background is from $\bar{q}q \to WW, WZ, ZZ$. Since $\mathcal{M}(\bar{q}q \to WW) = 0(g^2)$ while $\mathcal{M}(qq \to qqWW) = 0(g^2\lambda_{SB})$, we expect a discernible signal if and only if $\mathcal{L}_{SB}$ is strongly interacting, $\lambda_{SB} = 0(1)$. The signal may occur in W^+W^- and ZZ, as for the standard Higgs boson, but also more generally in $W^{\pm}Z$ and even W^+W^+ and W^-W^- ("I" = 1 and 2 channels).

To compute the expected yields we must convolute the luminosity distribution functions to find q and $\bar{q}$ beams in the incident protons with the luminosity distribution to find longitudinally polarized gauge bosons in the incident q's or $\bar{q}$'s, convoluted finally with the $2 \to 2$ scattering cross section of the longitudinally polarized gauge bosons:

$$\sigma(pp \to Z_L Z_L + \cdots) = \int_{\tau} \frac{\partial \mathcal{L}}{\partial \tau}\bigg|_{qq/pp} \cdot \int_x \sum_{V_L} \frac{\partial \mathcal{L}}{\partial x}\bigg|_{V_L V_L/qq} \cdot \sigma(V_L V_L \to Z_L Z_L) \quad (26)$$

where $\tau = s_{qq}/s_{pp}$ and $x = s_{ZZ}/s_{qq}$. In the effective W approximation the luminosity function for $W_L^+ W_L^-$ pairs is [18]

$$\frac{\partial \mathcal{L}}{\partial x}\bigg|_{W_L W_L/qq} = \frac{\alpha^2}{16\pi \sin^{.4}\theta_W} \frac{1}{x}\left[(1+x)\ln\frac{1}{x} + 2x - 2\right] \quad (27)$$

The x dependence of this function has the strong dependence on the available phase space that will be familiar to practitioners of two photon physics at e^+e^- colliders. Doubling the energy of the incident quarks from $\sqrt{s_{qq}} = 2\sqrt{s_{WW}}$ to $\sqrt{s_{qq}} = 4\sqrt{s_{WW}}$ increases the $W_L W_L$ luminosity by a factor 20! This accounts for the great sensitivity of the signals discussed below to the total collider energy.

	20 TeV	40 TeV
ZZ	150, 90, 200	370, 470, 890
W^+W^-	660, 120, 410	1600, 630, 1790
$W^{\pm}Z$	290, 120	670, 670
$W^+W^+ + W^-W^-$	0, 200	0, 1200

Table 1:   Yields in events per $10^4 pb^{-1}$ for 20 and 40 TeV pp colliders, taken from ref. [8] (with correction discussed in the text). Cuts are $M_{VV} > 1.0$ TeV and $|y_V| < 1.5$ except for $W^+W^+ + W^-W^-$ for which the rapidity cut is relaxed to $|y_V| < 4.0$. In each entry the first number is the $\bar{q}q$ annihilation background, the second is the conservative strong interaction model, and the third is the 1 TeV Higgs boson.

Yields are presented in Table 1 for the rapidity cut $|y_V| < 1.5$ and invariant mass cut $m_{VV} > 1.0$ TeV. These cuts are needed to see the ZZ, $W^\pm Z$, and W^+W^- signals over the $\bar{q}q$ annihilation backgrounds. Results are shown for 20 and 40 TeV colliders with $10^4 pb.^{-1}$ integrated luminosity. In each box of the table the first number denotes the $\bar{q}q$ background, the second is the conservative model, and the third, where present, is the 1 TeV standard model Higgs boson. The values are taken from reference [15] except that an error in the Higgs boson yield has been corrected (reducing those yields by $\sim 20\%$ relative to reference [15]). For like–charged WW pairs the rapidity cut in relaxed to $|y_W| < 4$ since there is no $\bar{q}q$ background, the leading background from single gluon exchange being perhaps ~ 5 times smaller than the $\bar{q}q$ backgrounds in the other channels.[19]

Notice the uncharacteristically sharp dependence on the machine energy, 20 TeV versus 40 TeV. This is simply because the signal is at the edge of phase space. Not only the signal but also the signal:background ratio suffer at lower energy. The greater sensitivity of the signal than the background reflects the four body phase space of the signal, $qq \to qqWW$, compared to the two body phase space of the background $\bar{q}q \to WW$. To compensate for lower energy, luminosity would need to be increased beyond what would be needed just to equal the signal of a higher energy machine.

Table 1 is chiefly a theoretical exercise since it does not necessarily correspond to experimentally implementable signals. We need to consider how the gauge bosons decay and are detected. This has been done more completely for the 1 TeV Higgs than for the conservative model, though in both cases much work remains to be done.

The cleanest channel, $ZZ \to e^+e^-/\mu^+\mu^- + e^+e^-/\mu^+\mu^-$, has a small branching ratio, $B = 3.6 \times 10^{-3}$. For the 1 TeV Higgs with $|y_Z| < 1.5$ and $m_{ZZ} > 0.9$ TeV this gives a signal of 4 events over a background of 1 for $\sqrt{s} = 40$ TeV. A more promising leptonic channel is[15,20] $ZZ \to e^+e^-/\mu^+\mu^- + \bar{\nu}\nu$ with $B \sim .026$, analogous to observing $W \to e/\mu + \nu$ at the SPS collider. The signal is defined by 1) $Z \to e^+e^-/\mu^+\mu^-$ at large p_T with central rapidity, 2) large missing p_T, and 3) no hot jet activity in order to veto the background from $W+$ jet. The latter especially must be studied with a Monte Carlo but is likely to be both clean and efficient. For the 1 TeV Higgs, Cahn and I find[20] that cuts of $|y_{e^+e^-}| < 1.5$

and $p_T(e^+e^-) > .45$ TeV give a signal of 27 events over a $\bar{q}q$ background of 8 for 40 TeV (10 standard deviations) compared to 6 over 3 for 20 TeV. (These values correct an error found in ref. [20] by M. Golden.) The $\bar{q}q$ backgrounds are under adequate theoretical control and will in any case be measured at the SSC.

The ZZ yields from the conservative, resonance–free strong interaction model of section 3 are smaller by about a factor 2 than for the 1 TeV Higgs boson. In the channel $ZZ \to e^+e^-/\mu^+\mu^- + e^+e^-/\mu^+\mu^-$ with $|y_Z| < 1.5$ and $m_{ZZ} > 1.0$ TeV we find a signal/background of only 2/1 events for $\sqrt{s} = 40$ TeV and only 0.4/0.7 for $\sqrt{s} = 20$ TeV. For the channel $ZZ \to e^+e^-/\mu^+\mu^- + \bar{\nu}\nu$ with $|y_Z| < 1.5$ and $p_T > 0.45$ TeV for the observed Z we find signal/background of 15/8 events at 40 TeV (for 5σ statistical significance) compared to 3/3 events at 20 TeV.

A straightforward leptonic channel is $W^\pm Z \to e\nu/\mu\nu + e^+e^-/\mu^+\mu^-$ with branching ratio 0.011. Though there is no WZ signal in the standard Higgs model, this channel may more generally exhibit important strong interaction effects. Defining the signal by $|y_{W,Z}| < 1.5$ and $m_{WZ} > 1.0$ TeV, the yield at $\sqrt{s} = 40$ TeV from the conservative model is $7\frac{1}{2}$ events over a $\bar{q}q \to WZ$ background of 3. If $W \to \tau\nu$ is also included, signal and background are increased by 50%. Dramatic resonance effects might arise in this channel. For instance, in the $N = 4$ technicolor model, the techni–rho meson is predicted at 1.8 TeV, and its charged states will decay predominantly to $W_L^\pm Z_L$. It is produced both by $\bar{q}q$ annihilation and WZ fusion. For the 40 TeV collider we find[15] 12 spectacular events over a $\bar{q}q \to WZ$ background of only 1 event in the leptonic channel with only e's and μ's, increased to 18 over 1.5 if $W \to \tau\nu$ is included.

The like–charged WW signal of the conservative model will also give rise to striking leptonic events of the form $W^+W^+/W^-W^- \to \ell^+\ell^+/\ell^-\ell^- + \nu\nu$, with no background from $\bar{q}q$ annihilation. For $\ell = e$ or μ the decay branching ratio is $B = .028$ so that corresponding to the cuts $|y_W| < 4$ and $m_{WW} > 1.0$ TeV there would be 33 events. Of course these cuts are not experimentally implementable in this decay channel. An experimentally meaningful set of cuts remains to be defined, but I expect cuts can be found that will lead to a signal of a few tens of events. Notice that for this channel we must be able to determine the sign of the charge for muons and electrons. If we could only measure the muon charge the yields would drop by a factor 4.

The leptonic decay channels discussed above are experimentally clean but suffer from small branching ratios. There are larger yields to be had if we can detect the hadronic decays of W and Z, however we then encounter formidable QCD backgrounds. For instance, the "mixed" channel $H \to WW \to u\bar{d}/c\bar{s} + e\bar{\nu}/\mu\bar{\nu}$ has a QCD background from $pp \to Wjj$ where the dijet fakes a $W \to \bar{q}q$ decay. Assuming a (perhaps optimistic) 5% resolution on the jj mass, the QCD background is two orders of magnitude larger than the signal.[21] Since the QCD background is dominated by gluon jets, it would be very helpful if we could learn to distinguish quark and gluon jets.

So far two approaches have been taken to the problem of winnowing this mixed decay signal from the the enormous QCD background. One approach[22] is to apply p_T cuts to the observed jets, using the tendency of the longitudinally polarized W's to decay into jets that are transverse to the W line of flight, unlike both the QCD dijets and the transversely polarized W's which both tend to give jets along the W line of flight. Extrapolating the results of ref. [22] for $m_H = 0.8$ TeV to $m_H = 1.0$ TeV suggests a possible signal of ~ 400 events over a background of ~ 400. This would be a formidable result if it can actually be accomplished in the laboratory, since it corresponds to reducing the background by almost three orders of magnitude while diminishing the signal by less than one order of magnitude.

A second approach to the mixed modes is to borrow a trick from two photon physics at e^+e^- colliders where detecting a final state $e^\pm$ in the forward direction is a powerful way to isolate a clean sample of two photon events. The analogous idea[23] is to tag the forward jets that occur in WW fusion, $qq \to WW + q + q$, with transverse momentum of 50–100 GeV because of the W mass. Of course this approach also has its own QCD backgrounds, from processes with a dijet faking a W and one or two forward jets faking the tagged quark or quarks. Estimating (but not yet calculating) the QCD background, the results of ref. [23] suggest that for the 1 TeV Higgs boson a signal of a few hundred events over a comparable background is probably attainable.

All this work, for both leptonic and mixed decay modes, is still at a preliminary state. Much remains to be done to refine the cuts that define the signals at the parton level and to determine how well they can be implemented by using Monte Carlo methods with detector simulations.

5. CONCLUSION

With the SSC design parameters, $\sqrt{s} = 40$ TeV and $\mathcal{L} = 10^{33}$cm^{-2}s^{-1}, we are assured of the capability to see the signal of a strongly interacting symmetry breaking sector. The SSC is strategically positioned in that it is close to being the minimal machine about which this statement can be made. If we do see signs of TeV scale, strong interaction, symmetry breaking physics, then (hard though it may be to imagine now) we are certainly going to think about a next generation facility that will allow more detailed studies.

Absence of the TeV-scale gauge boson pair signal would indicate that $\mathcal{L}_{SB}$ is weakly interacting with quanta well below the TeV scale. In that case the quanta of $\mathcal{L}_{SB}$ are copiously produced and in most physics scenarios it would be possible to study them in some detail.

As shown by the conservative strong interaction model, which assumes no resonances, we may expect the strong interaction signal to be visible even in the (unlikely) event that $M_{SB} >> 1$ TeV so that the strongly interacting quanta of $\mathcal{L}_{SB}$ are too heavy to produce at a 40 TeV collider. In fact, in that case the low energy theorems would enjoy the maximum possible range of validity. But a more likely prospect is for resonances to occur with the onset of strong scattering as happens in hadron physics. In that case we would observe resonances in the gauge boson pair cross section at a mass scale $M_{SB} \lesssim 0(2$ TeV$)$.

Epilogue: A Better World for our Children

The subprocess cross section for $W_L W_L$ scattering obtained by extrapolating the low energy theorems into the TeV energy range where they saturate unitarity is

$$\hat{\sigma}_{W_L W_L}\big|_{1-2\,\text{TeV}} \sim \frac{1}{2v^2} \sim O(1nb.) \tag{28}$$

which at the SSC translates into an observable cross sections of order a few tenths of a picobarn. If $\mathcal{L}_{SB}$ is strongly coupled then at high energies we may expect geometrical total cross sections with the size scale set by the W boson Compton wave length, just as the pion Compton wave length fixes the scale of hadron cross sections. Then we would have

$$\hat{\sigma}_{W_L W_L}\big|_{>>1\,\text{TeV}} \sim \pi R^2 \sim \frac{\pi}{M_W^2} \sim O(100nb), \tag{29}$$

two orders of magnitude larger than eq. (28). At low energy, eq. (28), the cross section is dominated by just the lowest partial waves, whereas at high energy many

56

partial waves contribute, up to $\ell = pR$. If the SSC probe of the TeV scale reveals strongly coupled structure from the symmetry breaking sector, then our children or grandchildren will use the SSC tunnel and high temperature superconducting magnets for a multi-hundred TeV accelerator that will be for the physics of symmetry breaking what the PS and AGS have been for the physics of hadrons.

Acknowledgments

I wish to thank Mitchell Golden, Robert Cahn, and Mary Gaillard for helpful discussions. This work was supported by the Director, Office of Energy Research, Office of High Energy and Nuclear Physics, Division of High Energy Physics of the U.S. Department of Energy under Contract DE-AC03-76SF00098.

References

1. M. Chanowitz, LBL-24878, 1988 (to be published in Ann. Rev. Nucl. Part. Sci.).

2. B. Holdom, Phys. Rev. D24 (1980) 1441; W. Bardeen et al., Phys. Rev. Lett. 56 (1986) 1230; K. Yamawaki et al., Phys. Rev. Lett. 56 (1986) 1335; T. Akiba and T. Yanagida, Phys. Lett. 169B (1980) 432; T. Appelquist et al., Phys. Rev. D35 (1987) 774; D36 (1987) 568.

3. M. Bando et al., Phys. Rev. Lett. 54, (1985) 1215; Nucl. Phys. B259 (1985) 493; M. Abud et al., Phys. Lett. 159B (1985) 185; P.Q. Hung, Phys. Lett. 168B (1986) 253; R. Casalbuoni et al., Nucl. Phys. B282 (1987) 335; H. Stremnitzer, U. Maryland report 86-20 (1986).

4. H. Georgi et al., Phys. Lett. B145 (1984) 216; B143 (1984) 152; B136 (1984) 187; Nucl. Phys. B254 (1985) 299.

5. R. Chivukula and H. Georgi, Phys. Lett. 188B (1987) 99; R. Chivukula, H. Georgi, and L. Randall, Nucl. Phys. B292 (1987) 93; R. Chivukula and H. Georgi, Phys. Rev. D, in press.

6. S. Adler and R. Dashen, Current Algebras, (W.A. Benjamin, Inc., N.Y., 1968).

7. T.D. Lee and C.N. Yang, Phys. Rev. Lett. 4 (1960) 307; B. Ioffe, L. Okun, and A. Rudik, Sov. Phys. JETP Lett. 20 (1965) 1281.

8. S. Weinberg, Phys. Rev. Lett. 19 (1967) 1264; A. Salam in "Elementary Particle Physics" (ed. N. Svartholm, Almqvist and Wiksells, Stockholm, 1968) p. 367.

9. M. Chanowitz, LBL-21973, to be published in the Proceedings of the XXIII International Conference on High Energy Physics, Berkeley, 1986.

10. S. Weinberg, Phys. Rev. Lett. 17 (1966) 616.

11. M. Chanowitz, H. Georgi and M. Golden, Phys. Rev. Lett. 57 (1986) 2344 and manuscript in preparation.

12. R. Cahn and S. Dawson, Phys. Lett. 136B (1984) 196; E 138B (1984) 464. See also D. Jones and S. Petcov, Phys. Lett. 84B (1979) 440; and Z. Hioki, S. Midorikawa, and H. Nishiura, Prog. Theor. Phys. 69 (1983) 1484. For the effective W approximation see M. Chanowitz and M. Gaillard, Phys. Lett. 136B (1984) 196; S. Dawson, Nucl. Phys. B29 (1985) 42; G. Kane, W. Repko, and W. Rolnick, Phys. Lett. 148B (1984) 367. For a study which does not rely on the effective W approximation, see G. Altarelli, B. Mele, and F. Pitolli, U. Roma preprint 531, 1986.

13. B. Cox and F. Gilman, Proc. Snowmass 1984 (ed. R. Donaldson) p. 87.

14. M. Bento and C. H. Llewellyn–Smith, Oxford U. preprint, 1986. See also G. Kane and J. Scanio, CERN TH 4532/86 and G. Altarelli et al., op. cit.

15. M. Chanowitz and M. Gaillard, Nucl. Phys. B261 (1985) 379.

16. M. Einhorn, Nucl. Phys. B246 (1984) 75; R. Casalbuoni, D. Dominici, and R. Gatto, Phys. Lett. 147B (1984) 419.

17. M. Duncan, G. Kane, and W. Repko, Nucl. Phys. B272 (1986) 517.

18. See papers cited in ref. [5] by Chanowitz and Gaillard; Dawson; and Kane, Repko, and Rolnick.

19. J. Gunion et al., Theoretical Report of $W/Z/$Higgs SSC Working Group, UCB-86-39, to be published in Proc. Snowmass 1986.

20. M. Chanowitz and R. Cahn, Phys. Rev. Lett. 56 (1986) 1327.

21. W. Stirling, R. Kleiss, and S. Ellis, Phys. Lett. 163B (1985) 261; J. Gunion, A. Kunzst, and M. Soldate, Phys. Lett. 163B(1985) 389.

22. J. Gunion and M. Soldate, Phys. Rev. D34 (1986) 826.

23. R. Cahn, S. Ellis, R. Kleiss, and W. Stirling, LBL–2164 (1986), to be published in Phys. Rev. D35, 1626 (1987).

58

AN ULTRAVIOLET FIXED POINT IN QUANTUM ELECTRODYNAMICS

Elbio Dagotto[†]

University of Illinois at Urbana-Champaign
Department of Physics
Loomis Laboratory
1110 W Green St., Urbana, Il 61801

ABSTRACT

The present status of numerical simulations of noncompact QED in 4 dimensions is reported. In the quenched approximation we measured $< \bar{\psi}\psi >$ near the chiral phase transition between strong and weak coupling. A good agreement is found with analytic predictions based on the "collapse of the wave function" ideas near the scaling region. Preliminary results with dynamical fermions are also discussed.

QED is considered a very successful quantum field theory. An excellent agreement has been found between the perturbative predictions of the theory and some experimental results. However, conceptually QED is still a poorly understood theory. For example it is known that the perturbative expansion is asymptotic, i.e., if we would be able to add many more terms to the present day perturbative calculations, the results would eventually diverge.

There is another unsatisfactory feature of QED: the Landau ghost problem.[1] Take as an approximation to the full photon propagator the sum of Feynman diagrams containing disconnected fermionic loops as shown in Fig. 1. In 3+1 dimensions each loop diverges like $ln(\Lambda)$ where Λ is a cutoff in momentum space. The diagrams of Fig. 1 can be summed and from them we can evaluate the renormalized charge of QED (e_R) as a function of the bare charge (e_0), the cutoff and a fermionic

---

† Address after September 1, 1988: Institute for Theoretical Physics, Univ. of California at Santa Barbara, Santa Barbara, CA 93106.

mass m used as an infrared regulator. The result is,

$$e_R^2 = \frac{e_0^2}{1 + \frac{e_0^2 N_f}{3\pi} \ln(\frac{\Lambda^2}{m^2})},$$ (1)

where N_f is the number of fermionic species in the theory (leptons) with mass m.

Figure 1. Approximation to the full photon propagator considered in eq. (1).

From eq. (1) it is clear that $e_R^2 \to 0$ when the cutoff is removed ($\Lambda \to \infty, e_0 = fixed$). In other words, when a test charge is introduced in the vacuum of QED, pair creation around the charge totally screens it and the resulting charge observed at large distances is zero. That means that QED is "trivial." Formally we need a cutoff in the theory to make it nontrivial (the cutoff may be related with the next scale of new physics, probably a grand unified theory). It is important to remark that the "one-loop" approximation shown in Fig. 1 is exact at large momenta[2] so the "triviality" problem is not an artifact of the particular subset of diagrams I considered in that figure.

Away from the weak coupling domain, is it possible to begin with the bare Lagrangian of QED, remove the cutoff and still get nontrivial physics? Or the Landau ghost problem survives the strong coupling limit of QED making the theory formally useless without a finite Λ ? These questions were the motivation of our project (done in collaboration with J.Kogut and A. Kocić). It is clear that we will need a nonperturbative technique to answer them so we use the lattice approach.

Our hope is that the beta function of QED may have the form depicted in Fig. 2 where at the origin we have an infrared stable fixed point leading to the triviality problem while at some finite value of the bare charge, say e_c^2, another zero exists. This new fixed point would be ultraviolet stable and we may obtain a nontrivial theory around it in the continuum limit. Such a point will show up in a lattice simulation as a second order phase transition. If we find this transition, then the next step would be the study of the neighborhood of the critical region looking for nontrivial scaling near the continuum limit. Note that the existence of

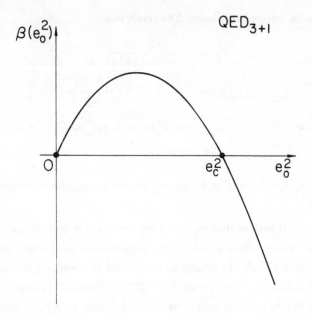

Figure 2. Tentative beta function for QED in 3+1 dimensions.

an ultraviolet fixed point in QED has been previously suggested by using Schwinger-Dyson equations in the quenched ladder approximation.[3,4,5]

First we tried to find the ultraviolet fixed point in the **compact** formulation of QED. In this case we know that a deconfining phase transition exists in the pure gauge limit. Working with lattices of 6^4 and 8^4 sites and a fermionic mass of 0.10 (lattice units) we found two clearly different regions separating a phase where chiral symmetry is broken from a symmetric (weak coupling) phase. However, the phase transition is of first order and we can not recover a continuum limit around it since the correlation length is always finite. We also tried with a mixed action but again the transition was discontinuous even in places where the order is controversial in the pure gauge limit. In fact the gap between the two metastable states at the phase transition increases with N_f. These results have been discussed in previous articles.[6] We do not know the origin of this first order transition (an intuitive understanding of it would be welcomed).

As a second attempt we tried with the **noncompact** version of QED and in this case we found better results.[7] The action is given by

$$S = -\frac{\beta}{2} \sum_p (\theta_p^2) + \sum_{x,y} \bar{\psi}_x M_{x,y} \psi_y, \qquad (2)$$

where

$$M_{x,y} = \frac{1}{2} \sum_\mu \eta_{x,\mu} [e^{i\theta_{x,\mu}} \delta_{y,x+\mu} - e^{-i\theta_{x,\mu}} \delta_{y,x-\mu}] + m\delta_{x,y}. \qquad (3)$$

We used staggered fermions. The notation is standard. Note that $\theta_{x,\mu}$ can take values between $-\infty$ and $+\infty$ and not only between $(0, 2\pi)$. However, we do not need to fix a gauge: if during the simulation a variable takes a big value, we can always reduce it by a gauge transformation.

The noncompact action eq. (2) is gauge invariant (we are working with an abelian group). In the absence of fermions it is a free theory (not like compact QED that has a confining phase) and it is free from monopoles. We first studied this model using lattices of 6^4 and 8^4 sites. In the massless limit we found numerically that there is a continuous phase transition at a critical value $\beta_c \sim 0.20$ ($N_f = 4$). We observed no indications of metastability. These results have been reported in ref. [7].

After finding this second order phase transition we have to check that the continuum physics around it is nontrivial. In other words we should measure some quantities in the "scaling" region and verify that they scale with non-mean field critical exponents. The problem of QED in strong coupling, that makes it much more difficult than QCD in weak coupling, is that we do not have much intuitive control over the critical behavior of the theory and we can not evaluate exactly the scaling laws.

However, there are some analytic results in the literature concerning the ultraviolet fixed point of QED.[4,5] Let me begin with a simple quantum mechanical (QM) example and then I will go back to QED. Consider the Dirac equation in 3+1 dimensions with a Coulomb potential as shown in Fig. 3. We have introduced a cutoff $r_0 \ll 1$ to prevent possible problems at short distances. It is well known that this model exhibits a "collapse" in strong coupling:[8] when the coupling constant exceeds a critical value ($\alpha_c = 1$), the ground state energy falls to $-\infty$ and the ground state wave function develops an infinite number of zeroes. So in principle the model does not make sense for $\alpha > \alpha_c$. However, there is a solution for this problem:

Consider the ground state wave function for small r (but bigger than r_0) and for $\alpha > \alpha_c$. It is given by

$$\psi(r) \sim \frac{1}{r\sqrt{\alpha^2 - 1}} \sin\left[\sqrt{\alpha^2 - 1}\ \ln(\frac{1}{r|\epsilon|})\right], \tag{4}$$

where ϵ is the ground state energy. This function has an oscillatory behavior. But suppose that we keep ϵ fixed to an arbitrary value such that

$$\epsilon \sim -\frac{1}{r_0}\exp(-\frac{\pi}{\sqrt{\alpha^2 - 1}}), \tag{5}$$

while we remove the cutoff ($r_0 \to 0$) and we approach the critical point from the strong coupling side ($\alpha \to \alpha_c^+$). In that limit eq. (4) becomes

$$\psi(r) \sim \frac{1}{r}\ln(\frac{1}{r|\epsilon|}), \tag{6}$$

which has no zeroes and is therefore a reasonable ground state. The wave function eq. (6) has a finite probability at the origin so e^- and e^+ can collide at $r = 0$.

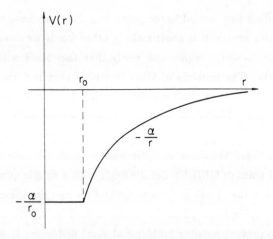

Figure 3. Coulomb potential used in the Dirac equation.

Note that eq. (5) represents a renormalization of charge. So even in a QM problem (in principle only a few body problem) we can have renormalization of bare parameters due to the collapse phenomena (which is another source of infinities besides the standard ones coming from perturbation theory). This renormalization of charge also tell us that the beta function of this example is nonzero for $\alpha > \alpha_c$.

This concept of "collapse of the wave function" in QM can be extended to quantum field theory using the self-consistent Schwinger-Dyson equations. For simplicity let us work in the quenched ladder approximation. We can write a (truncated) equation for the fermionic self-energy $\Sigma(k)$ as shown in Fig. 4 where the continuous line with a dot represents the full propagator, without the dot the free propagator and the wavy line is a free photon. In this approximation there are no vertex renormalization nor photon propagator corrections. The equation represented by Fig. 4 was solved in refs. [4,5]. It was found that there is a critical point at $\alpha_c = \pi/3$ beyond which chiral symmetry is spontaneously broken. An analysis of the Bethe-Salpeter wave function shows again an infinite number of zeroes for $\alpha > \alpha_c$. As before we can fix this problem by a renormalization of charge when we approach α_c from strong coupling keeping fixed some physical mass. The relation equivalent to eq. (5) is in here,

$$m_F \sim \Lambda \quad \exp(-\frac{\pi}{\sqrt{\frac{\alpha}{\alpha_c} - 1}}) \tag{7}$$

where m_F is a fermionic mass. On the other hand, very deep in strong coupling the dynamical mass scales like $m_F \sim \Lambda\sqrt{\alpha - \alpha_c}$, i.e., with mean field exponents.[9]

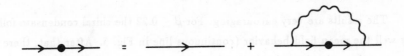

Figure 4. Graphical representation of the Schwinger-Dyson equation considered in the text.

So the analytic prediction in the quenched approximation is that $< \bar{\psi}\psi >$ is nonzero for $\alpha > \alpha_c$ showing a mean field behavior away form the critical point that changes to an essential singularity behavior near the critical point when the coherence length is much bigger than the lattice spacing.[10]

To check these ideas now I present our more recent numerical results.[9] In Fig. 5 we plot $< \bar{\psi}\psi >$ vs. $\beta(= 1/e_0^2)$ for a 16x8^3 lattice in the quenched

64

approximation. For the simulation we used the hybrid technique with noisy fermions[11] with a time step $dt = 0.02$. Each point in Fig. 5 represents a linear extrapolation to $m = 0$ of the data obtained with $m = 0.02$ and 0.04 (using 125000 iterations for each one of these points).

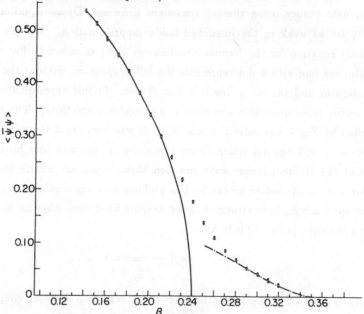

Figure 5. $< \bar{\psi}\psi >$ vs. β extrapolated to $m = 0$ (quenched approximation).

The results are very encouraging. For $\beta \leq 0.22$ the chiral condensate follows very well the mean field behavior (continuous line in Fig. 5. After that, there is a crossover where the data clearly depart from mean field and finally they converge to an essential singularity behavior (dashed line) given by

$$< \bar{\psi}\psi > \sim \exp(\frac{-b}{\sqrt{\beta_c - \beta}})$$ (8)

where $\beta_c \sim 0.36$ and b is an unimportant constant. In a future publication[12] we will present all our numerical data for the quenched approximation.

What happens when dynamical fermions are introduced in the simulation? In principle we expect them to be very important. Take for example the weak coupling region: when we are allowed to create $e^- e^+$ loops, then the Landau ghost problem arises. In the quenched approximation there is no such a problem but a nonzero N_f (no matter how small) brings it back, at least if we trust eq. (1). Numerically the introduction of fermions require an order of magnitude more computing time than at $N_f = 0$ so we did only between 10000 and 50000 iterations per point. The results for $N_f = 2$ and 4 are shown in Fig. 6 on a 8^4 lattice again using $m = 0.02$ and 0.04 for a linear extrapolation to $m = 0$. For $N_f = 4$ the steepness of the curve suggests that the mean field result may become exact and the continuum limit could be trivial (the little "tail" in Fig. 6 for $N_f = 4$ can hardly be distinguished from a finite size effect). For $N_f = 2$ we obtained an intermediate result between those at $N_f = 0$ and 4. Although fits similar to eq. (8) can be made here, more work is needed to clearly prove that the continuum limit is nontrivial (that work is in progress).

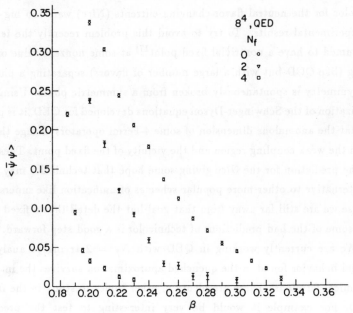

Figure 6. $N_f = 0, 2$ and 4 data of $< \bar\psi\psi >$ vs. β

Do we expect some physical consequences related with the existence of the strong coupling phase of QED? This is still controversial. One possible application comes from the narrow e^-e^+ peaks found in heavy ion collisions[13] that look like coming from the decay of e^-e^+ bound states. It has been speculated[14] that these bound states may be produced by a phase transition to the strong coupling phase of QED triggered by the large charge of the two heavy ions during the collision. However, preliminary numerical results do not support this scenario[15] although probably the external electromagnetic configuration considered there (a plain Coulomb field) is too naive to model the collision.[16]

Another possible application of the new fixed point is related with technicolor theories. These models try to avoid the use of Higgs fields to generate the masses for the intermediate gauge bosons. To do that a new "technigroup" is introduced. The fermions $T, \bar{T}$ associated with the new group condense (like quarks in QCD) and $< \bar{T}T >$ replaces the nonzero mean value of the Higgs field (unitary gauge) in the electroweak theory. For simplicity the technigroup was originally considered as an asymptotically free group. But it was quickly realized that the predictions of technicolor for the neutral flavor-changing currents (Nfcc) were too big compared with experimental results. To try to avoid this problem recently the technigroup was assumed to have a nontrivial fixed point[17] at some nonzero value of the bare coupling (like QCD but with a large number of flavors) separating a phase where chiral symmetry is spontaneously broken from a symmetric phase. Using a simple generalization of the Schwinger-Dyson equations developed for QED, it is possible to show that the anomalous dimension of some 4-Fermi operators change their values between the weak coupling region and the vicinity of the fixed point. These changes affect the prediction for the Nfcc giving some hope that technicolor may become a good alternative to other more popular schemes of unification like supersymmetry. Of course we are still far away from that goal but the detail that a fixed point can change some of the bad predictions of technicolor is a good step forward.[18]

We are currently working in QED with $N_f = 2$ trying to analyze if the nontrivial behavior found in the quenched approximation survives the introduction of dynamical fermions. There are plenty of other topics to study in the framework of QED. For example it would be very interesting to test the prediction of refs. [4,17] regarding the existence of renormalizable 4-Fermi operators in the effective Lagrangian of the theory near the fixed point (remember that in the QM example discussed above the probability at the origin of the renormalized wave function was nonzero and that may induce a "contact" term). Also, it would be nice to generalize the collapse ideas to other models like QED with Higgs fields[19] and Yukawa couplings, and analyze its possible relevance for the electroweak theory. The author is supported by the National Science Foundation grant PHY87-01775. The computer simulations were done on the CRAY X-MP/48 of the National Center for Supercomputing Applications at Urbana. I thank A. Kocić, J. Kogut and A. Moreo for useful discussions.

References

1. L. D. Landau, Niels Bohr and the Development of Physics, eds. W. Pauli (Mc Graw Hill, NY, 1955).

2. L. D. Landau and I. Ya. Pomeranchuk, Dokl. Akad. Nauk., 102, (1955) 489.
E. S. Fradkin, Sov. Phys. JETP, 28, (1955) 750.
L. D. Landau, A. Abrikosov and L. Halatnikov, Supl. Nuovo Cimento, 1, (1956) 80.

3. K. Johnson, M. Baker and R. Willey, Phys. Rev., 136B, (1964) 111.
S. Adler, Phys. Rev., D5, (1972) 3021.

4. P. Fomin, V. Gusynin, V. Miransky and Yu. Sitenko, Riv. Nuovo Cimento, 6, (1983) 1.
V. Miransky, IL. Nuovo Cimento, 90A, (1985) 149.

5. C. N. Leung, S. T. Love and W. A. Bardeen, Nucl. Phys., B273, (1986) 649.

6. E. Dagotto and J. Kogut, Nucl. Phys., B, (1988).
J. Kogut and E. Dagotto, Phys. Rev. Lett., 59, (1987) 617.

7. J. Kogut, E. Dagotto and A. Kocić, Phys. Rev. Lett., 60, (1988) 772.

8. See for example: L. Landau and E. Lifshitz, Relativistic Quantum Theory (Pergamon Press, 1971) 110.

9. J. Kogut, E. Dagotto and A. Kocić, On the existence of quantum electrodynamics, submitted to Phys. Rev. Letters.

10. Note that the situation has many similarities with the XY model in 2 dimensions: in that case there is a theory (vortex condensation) that is not exact but that has explained many of the numerical results found in that model. In this case also the scaling behavior is given by an essential singularity.

11. S. Gottlieb et al., Phys. Rev., D35, (1987) 2531.
G. Batrouni et al., Phys. Rev., D32, (1985) 2736.

68

12. E. Dagotto, A. Kocić and J. Kogut, in preparation.

13. T. Cowan et al., Phys. Rev. Lett., 56, (1986) 444.

14. L. Celenza et al., Phys. Rev. Lett., 57, (1986) 55.
    D. Caldi and A. Chodos, Phys. Rev., D36, (1987) 2876.
    Y. Jack Ng and Y. Kikuchi, Phys. Rev., D36, (1987) 2880.
    C. W. Yong, Electrons and photons trapped in bags of abnormal QED vacuum, UCLA preprint.

15. E. Dagotto and H. Wyld, Phys. Lett., B205, (1988) 73.

16. Compact (pure gauge) QED in an external magnetic field has been studied by P. Damgaard and U. Heller NBI-HE-88-19. Again no numerical indication of a shift of β_c to weak coupling was found.

17. M. Bando, T. Morozumi, H. So and K. Yamawaki, Phys. Rev. Lett., 59, (1987) 389.

18. For nonabelian theories a noncompact formulation on a lattice breaks explicitly gauge invariance while a compact formulation with many fermions give first order transitions. This is a potential problem for numerical checks of the new technicolor ideas.

19. I-Hsiu Lee and R. Shrock, Nucl. Phys., B 290, (1987) 275.
    E. Dagotto and J. Kogut, Towards a non perturbative study of the strongly coupled standard model, to appear in Phys. Lett. B ILL-(TH)-88-9.

TRIVIALITY IN THE 1/N EXPANSION: (1) STRONGLY INTERACTING HIGGS BOSONS; (2) VERY HEAVY FERMIONS

Martin B. Einhorn

Department of Physics
University of Michigan
Ann Arbor, MI 48109-1120

ABSTRACT

We review manifestations of Triviality which appear in the analytic solutions to theories via $1/N$ expansions. We discuss two examples of continuum formulations with a momentum cutoff: (1) the O_{2N} Higgs-Goldstone model in the limit of a strongly interacting Higgs boson, relevant to the issue of the upper limit on the Higgs mass in the Standard Model, and (2) a $U_N \otimes U_N$ Yukawa theory in the limit of strong Yukawa coupling. We show that both are trivial, so that there is an upper limit to the Higgs mass and to fermion masses. In an Appendix, we express some cautious concern about recent results suggesting a perturbative upper limit to the Higgs mass, based on lattice simulations.

In this talk, I shall review some results obtained via a $1/N$ expansion characterizing triviality, first, in a strongly interacting Higgs system and, secondly, for a system involving strong Yukawa interactions (but only in the case of weak scalar self-coupling). In the context of these Proceedings, I can dispense with the motivation for considering such questions. Unlike most of the rest of this volume, I will work in the continuum limit, since the discussion is easier analytically and since this is the situation the lattice approach hopes to reproduce. However, the models illustrate analytically many of the results which have been found in lattice simulations. In an Appendix, I add a note of skepticism concerning the conclusions drawn from some of the lattice results obtained thus far on the mass of the Higgs

Boson[1] and indicate what issues must be resolved before such results can be taken seriously for the continuum theory.

1. STRONGLY COUPLED HIGGS

In the Standard Model (SM), the H mass m_H grows with the scalar self-coupling λ, in tree approximation, $m_H^2 = s\lambda v^2$, where the weak scale v is fixed by the Fermi Constant G_F. In the limit that the $SU_2 \otimes U_1$ gauge couplings, g_1^2 and g_2^2 may be neglected relative to the scalar self-coupling λ, the scalar sector of the SM becomes the O_4 Higgs-Goldstone Model. The equivalence theorem between the Higgs-Goldstone model and the scattering of longitudinal vector bosons at energies which are large compared to their mass, makes the relevance of the O_4-model to the SM precise.[1] Some time ago, I suggested[2] that, to obtain qualitative insight into the behavior of the $SU_2 \otimes U_1$ Higgs sector of the SM for a strongly interacting (i.e. nonperturbative) H, one might consider instead the behavior of the $SU_N \otimes U_1$ model. The idea was very much in the same spirit that SU_N is sometimes useful to gain insight into the dynamics of SU_3-color in QCD.[3] Just as the SM goes over to O_4, neglecting gauge interactions $SU_N \otimes U_1$ becomes O_{2N}.[2] This can easily be seen as follows: The $SU_N \otimes U_1$ Higgs model, characterized by a (complex) Higgs field ϕ_n $(n = 1, \ldots, 2N)$ in the fundamental representation $\underline{N}$ of SU_N, may be written in terms of $2N$ real fields $\pi_i(i = 1, \ldots, 2N)$ by $\phi_n \equiv \frac{-i}{\sqrt{2}} (\pi_{2n-1} - i\pi_{2n})$ so that the SM Lagrangian becomes

$$\mathcal{L} = \frac{1}{2} \left(\partial_\mu \vec{\pi}_0 \right)^2 - \frac{\lambda_0}{4} \left(\vec{\pi}_0^2 - v_0^2 \right)^2, \tag{1}$$

where the subscript 0 reminds us that these are all bare quantities. Some cutoff is implicitly assumed to regulate the divergences of perturbation theory. For our purposes, it is most natural to discuss renormalization in the Euclidean field theory and employ a simple momentum cutoff restricting all momenta to $p_E^2 < \Lambda^2$. This theory is renormalizable as a perturbation expansion in λ for fixed N. In the limit that $\lambda_0 \to \infty$, it becomes nonrenormalizable; in the latter form in terms of $\vec{\pi}_0$,

---

[1] While the weak vector bosons are almost universally denoted as $W^\pm$ and Z^0, there is no standard notation for the unique Higgs Boson of the minimal Standard Model. In this talk, I simply call it the H-Boson or just H.

[2] This factor of 2 in $2N$ presents a slight departure from the conventional reference to the O_N model.

it is called the nonlinear σ model. The linear and nonlinear versions have been studied for a long time in several different space-time dimensions both for various fixed values of N and, in the $N \to \infty$ limit, in a variety of contexts: spin systems [4] as well as in field theory[5,6] and, with the addition of fermions, as a model of chiral symmetry breaking consistent with current algebra.[7]

I shall now describe the solution of $\mathcal{L}$ using the 1/N approximation. The idea of the 1/N for fixed[3] λN is to hope that one can sum analytically to *all* orders in λN. One might then be able to characterize the theory in a systematic but nonperturbative way.[4] To leading order, this can be carried out analytically in the functional integral.[8][5] We find that the contributions to leading order are a geometric series involving powers of the one-loop graphs. To proceed, our first task is to discuss the renormalizability of the theory. The details of the solution are discussed in Ref. [2], a summary of which may be found in a recent paper.[9] For this reason and because of space limitations here, I shall just state the results.

To $O(1)$ in 1/N, a mass-independent renormalized coupling λN may be related to the bare coupling $\lambda_0 N$ according to

$$\frac{1}{\lambda N(M)} \equiv \frac{1}{\lambda_0 N} + \frac{1}{4\pi^2} \ln\left(\frac{\Lambda}{M}\right). \qquad (2)$$

This has precisely the same *form* as the one-loop approximation; indeed, to $O(1)$ in 1/N, the one-loop formula for the beta-function $\beta_{\lambda N}$ is exact. It is important, however, to recognize that the relation expressed by Eq. (2) is non-perturbative in the coupling (though perturbative in 1/N) and exact to all orders in $\lambda_0 N$. To this order in 1/N, there is no wave-function renormalization, finite or infinite, only mass renormalization. One can show that mass divergences may be removed by defining

$$\frac{-\mu(M)^2}{\lambda N(M)} \equiv \frac{-\mu^2}{\lambda_0 N} + \frac{\Lambda^2}{8\pi^2}. \qquad (3)$$

---

[3] We'll treat only the purely scalar model here. For the $SU_N \otimes U_1$, the gauge couplings g_N^2 and g_1^2 are also held fixed.

[4] In the presence of gauge fields, the expansion is appealing because it is gauge invariant.

[5] Although in some ways the functional approach is more elegant, I prefer to do it by identifying those Feynman diagrams contributing to leading order and summing them,[2] because this can in principle be carried out beyond leading order and for nonzero gauge couplings. In addition, in the end, to determine how formal, functional integral arguments survive renormalization, one has to calculate diagrams anyway.

Note the quadratic divergence in the mass renormalization, an indication of the problem of fine-tuning the bare parameters in order to give a sensible renormalized mass.[10] We shall pass over this issue for the purposes of this talk.[6]

The coupling λN is *trivial*, i.e., requiring $\lambda_0 N > 0$, then $\lambda N(M) \to 0$ as $\Lambda \to \infty$ for every $M < \Lambda$. The hypothesis that $\lambda_0 N > 0$ is sometimes rationalized by requiring that the bare theory have a classically stable ground state, but this is not compelling beyond perturbation theory since radiative corrections can (and apparently do) provide stability if $\lambda N > 0$. Therefore, I prefer another aspect of triviality which does not place any restriction on the bare coupling. Given the renormalized coupling $\lambda N(M)$ there is a momentum scale Λ_c at which $\lambda N(M)$ is singular. To leading order in $1/N$, Λ_c may be defined as $\lambda N(M) \equiv 4\pi^2/\ln(\Lambda_c/M)$. Thus, Λ_c is the nonperturbative analogue of the Landau pole.[7]

Another feature of the renormalized coupling is that λN displays *universality*, i.e., when we let $\lambda_0 N \to \infty$ for fixed Λ, we find that $\lambda N(M)$ remains finite. This is a behavior which has been clearly verified by lattice simulations. It is clearly redundant to characterize the theory in terms of both a bare coupling λ_0 and a cutoff Λ. In fact, requiring that the bare parameters vary with Λ in such a way that the renormalized parameters be unchanged is an alternative way of deriving the renormalization group, so starting with the nonlinear σ model should result in no loss of generality.

In the Higgs phase ($\mu^2 > 0$), we have spontaneous breaking, with the $\vec{\pi}$ field developing a VEV, say, $\langle \pi_{2N} \rangle = v \equiv \mu/\sqrt{\lambda}$. We have $2N - 1$ Goldstone bosons and a single Higgs boson associated with $\sigma \equiv \pi_{2N} - v$. The Feynman rules are depicted in Fig. 1a. The VEV v is $O(\sqrt{N})$, and one can show that $v/\sqrt{N}$ receives no radiative corrections in leading order. The radiative corrections to the H self-energy are depicted in Fig. 1b, a geometric series closely related to one for the

---

[6] In fact, I think the sensitivity to the bare mass parameter in lattice simulations provides additional non-perturbative support for this concept, but I've never seen this discussed.

[7] However, for fixed, finite N, this becomes a one-loop rather than exact result so that one cannot justify this discussion of the UV behavior.

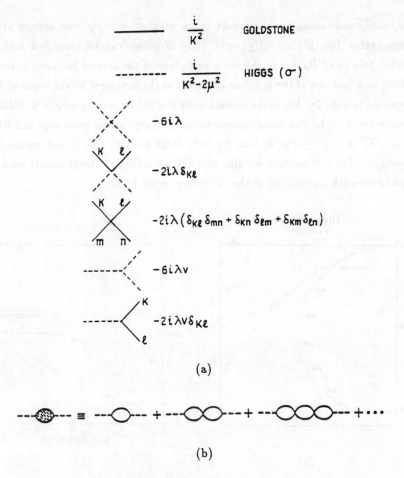

(a)

(b)

Figure 1. (a) Feynman Rules in the Higgs phase. (b) Leading corrections to the Higgs self-energy.

four-point function. The corresponding inverse H propagator is[8]

$$D_E^{-1} = p_E^2 + 2\lambda \left(\frac{\sqrt{p_E^2}}{e} \right) v^2,$$ (4)

that is, the propagator looks just like the tree approximation with the scale of the running coupling constant proportional to the momentum scale of interest.

---

[8] Here and subsequently, e does not denote a coupling constant but rather the natural base $\ln e \equiv 1$.

Analytically continuing to Minkowski space with $p^2 = -p_E^2$, one arrives at the H propagator $D_{\sigma\sigma}(p^2) \equiv -D_E(-p^2)$. The H-boson can be identified with the complex pole s_σ of $D_{\sigma\sigma}(s)$ in the lower half plane of the second Riemann sheet; its position as a function of the coupling strength, as characterized by the value of Λ_c/v is depicted in Fig. 2a. We make contact with the SM by setting $v/\sqrt{N} = 174 GeV$, its value for $N = 2$. The usual interpretation in terms of the mass m_H and width Γ_H is $\sqrt{s_\sigma} \equiv m_H - i\Gamma_H/2$, but, for very large widths, this is not particularly meaningful. For comparison, we also plot the perturbative (tree) results with the coupling strength normalized at the (arbitrary) scale $|\sqrt{s_\sigma}|$.[9]

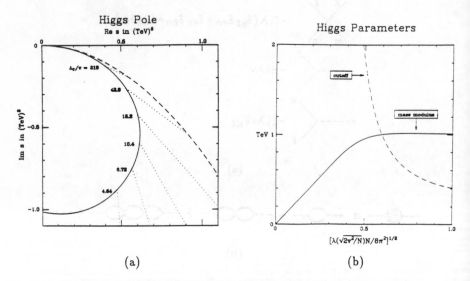

(a)   (b)

Figure 2. (a)Solid Line: Higgs pole position s_σ; Dashed line: perturbative Higgs pole position, normalized at $|s_\sigma|$. Dotted lines connect curves at points having a common value of Λ_c/v. (b)The Higgs mass modulus $|s_\sigma|$, and cutoff Λ_c as functions of the coupling strength.

Since momenta larger than Λ_c are beyond the reach of this field theory, one can determine the upper limit to the H "mass" by finding the crossover point

---

[9] It can be seen that the $1/N$ approximation is numerically rather poor for weak coupling. This can also be seen from the bubble graph, proportional to $2N + 7$ or the two-loop bubble proportional to $4N^2 + 20N + 9$. For $N = 2$ the terms neglected are bigger than the leading term. This approximation could be made to agree by keeping the same graphs but retaining the lower order terms of the bubble.

at which $|s_\sigma| = \Lambda_c^2$. This solution is depicted in Fig. 2b. This turns out to be $\Lambda_c \approx 1.8\pi v/\sqrt{N}$, which, with the aforementioned assignment of v as the weak scale, corresponds to an upper limit on the H modulus of about 1.0 TeV. From Fig. 2b it is easy to see that this upper limit is rather insensitive to weakening the crossover criterion. For example, if it should be that significant deviations from the SM already appear when $|s_\sigma| = \Lambda_c^2/4$, then the crossover is at $|\sqrt{s_\sigma}| = 0.86$ TeV instead. To be more precise requires a careful examination of the higher dimensional operators entering the effective Lagrangian and consideration of their effects in particular processes. It is perhaps fortuitous, but nevertheless striking, that the absolute upper limit determined here for the H modulus turns out to be numerically very nearly the same as the perturbative upper bound on the H mass![11]

It is also interesting to discuss the H propagator in coordinate space, especially since the H mass is determined in a number of lattice simulations by fitting this correlation function in Euclidean space. Before specializing to this model, let me say a few words about its generic behavior.[10] One can discuss the asymptotic behavior of the H propagator $\Delta_F(p^2)$ quite generally starting from the Källén-Lehmann representation.[12] In coordinate space, the propagator may be written as

$$\Delta_F(x) = \int_0^{\sim \Lambda_c^2} ds\, \rho(x)\, \Delta_F(x; s), \tag{5}$$

where the spectral function $\rho(s)$ is necessarily non-negative, vanishing below the lightest threshold to which the Higgs field is coupled. The upper limit of integration must be cutoff at a scale of $O(\Lambda_c^2)$ since we assume that a local field theory approximation is good only for smaller energy scales. $\Delta_F(x; s)$ is the free particle propagator associated with the particle of mass $\sqrt{s}$. In the Euclidean region $(x^2 \equiv -r^2 < 0)$, this is simply

$$i\Delta_F(x; s) \equiv \Delta_E(r; s) = \frac{\sqrt{s}}{4\pi^2 r} K_1\left(\sqrt{s}\, r\right). \tag{6}$$

For large r, $\Delta_E(r; s)$ behaves as $r^{-3/2} \exp(-\sqrt{s}\, r)$, so the true asymptotic behavior as $r \to \infty$ is determined by the behavior of $\rho(x)$ at the lightest threshold. In lowest order, the lower limit of integration is at $4m_e^2$ associated with the electron-positron

---

10 Much of the discussion that follows was developed in collaboration with D.N. Williams.[9]

state. In fact, for a very heavy H, this contribution as well as those due to other known fermions[11] is weak compared to the coupling to W^+W^- associated with the threshold at $4m_W^2$, and we shall neglect its contribution.

From the positivity of the spectral function, it can be shown that the Euclidean propagator $\Delta_E(r) \equiv i\Delta_F(x)$ is positive and monotonically decreasing with successive derivatives alternating in sign. Consequently, no matter what the dynamics, $\Delta_E(r)$ decreases smoothly without even so much as a wiggle! This may render the quality of a fit in a given model or approximation a poor indicator of the quality of the approximation.

As remarked earlier, it can happen that, because ρ is so much larger in a higher mass region, the behavior may be approximately described over an intermediate range of r by ignoring a lighter threshold. So it may be useful to record the contribution coming from a resonance. First, however, let us recall that because the propagator is an analytic function real in the Euclidean region associated with a pole at position s_σ, there is a mirror pole at s_σ^*. Therefore, we may write this contribution to the spectral function as

$$\rho_\sigma(s) = \frac{1}{2\pi i} \left(\frac{Z_\sigma^*}{s - s_\sigma^*} - \frac{Z_\sigma}{s - s_\sigma} \right). \qquad (7)$$

(This formula does not represent an analytically correct approximation to $\rho(s)$, but rather what we will mean by the pole approximation thereto. We suppose the "wave function renormalization constant" Z_σ is such that $\rho_\sigma > 0$.) One may then ask what this contribution gives for the coordinate space behavior when inserted into Eq. (5). As remarked earlier, the true asymptotic behavior will be determined by the threshold position, which, with the neglect of the vector boson masses, behaves as r^{-4}, as discussed below. However, it can be shown [9] that a pole approximation to the large r behavior yields

$$\Delta_E(r) \to \left(1 - \frac{2\Theta}{\pi} \right) \frac{1}{2(2\pi r)^{3/2}} |Z_\sigma| |s_\sigma|^{1/4} e^{-m_H r} \cos\left(\frac{\Gamma_H r}{2} - \frac{\Theta}{2} + \zeta \right), \qquad (8)$$

where we have defined $\sqrt{s_\sigma} \equiv m_H - i\Gamma_H/2 \equiv \sqrt{|s_\sigma|} \exp(-i\Theta)$, $Z_\sigma \equiv |Z_\sigma| \exp(i\zeta)$, and assumed $|s_\sigma| \gg 4M_W^2$. In principle, these formulae answer the question

---

[11] The inclusion of a very heavy fermion (as the top quark may very well be) requires a new treatment which we would expect to alter the results below quantitively but not qualitatively.

of how the H pole parameters are related to the Euclidean correlation function, demonstrating that a pole yields a behavior for the Euclidean Green's function which is oscillatory with an exponentially decaying envelope. Although Eq. (8) makes no assumption about the ratio Γ_H/m_H being small, a positive spectral function ρ_σ must, as noted above, produce a positive, monotonically decreasing propagator, so the approximations represented by Eqs. (8) can apply at most to a fraction of the distance out to their first zero. The question how the contribution from a resonance compares to the exact propagator, including the continuum "background" remains unanswered.

Returning to our solution for the propagator to leading order in $1/N$, one can show that the spectral function $\rho(s)$ may be written as

$$\rho(s) = \frac{16\pi^2 v^2/N}{\left(s \ln\left(\frac{e^2\Lambda^2}{s}\right) - 16\pi^2 v^2/N\right)^2 + \pi^2 s^2}. \tag{9}$$

The true asymptotic behavior is determined by the threshold behavior $\lim_{s\to 0}\rho(s) = \rho(0) = (16\pi^2 v^2/N)^{-1}$. The asymptotic behavior in coordinate space of the propagator in the Euclidean region may then be determined to be

$$\Delta_E(r) \to \frac{1}{\pi^2}\frac{\rho(0)}{r^4}. \tag{10}$$

Notice that this leading behavior is completely independent of the H mass parameters![12] In the real world where the W-boson has nonzero mass, this simple power behavior would be tempered by a factor of $\exp(-2M_W r)$. Nevertheless, it will decrease far less rapidly than might be expected on the basis of the H mass scale.

We[9] have done a certain amount of numerical work to determine whether the pole formula represents the actual behavior in this approximation. In the strongly coupled regime, as one might expect, the pole approximation Eq. (8) gives a rather poor description of the actual correlation function. On the other hand, replacing the actual spectral function by the pole approximation Eq. (7) yields a very good approximation, because, as it happens, this gives a value for $\rho_\sigma(0)$ which is not much smaller than the true value of $\rho(0)$. Thus, the actual correlation function does

---

[12] It is tempting to speculate that this is a general consequence of the infrared freedom of the scalar sector, but it remains to be proven.

not discriminate effectively between the underlying pole contribution to the spectral function and the continuum.

At worst, the $1/N$ expansion provides an analytically solvable model in which the concepts may be illustrated and explored and which, so far as is known, provides a qualitative but not quantitative guide to what can be expected of the SM. As we have heard at this Workshop, there have been a number of lattice simulations [13-16] suggesting that the upper limit to the m_H actually occurs in the perturbative domain with numbers in the range $550 \ GeV < (m_H)_{\max} < 750 \ GeV$. The main *qualitative* difference between these results and ours is that our upper bound occurs in the strong coupling region $850 \ GeV < \left(\sqrt{|s_\sigma|} \right)_{\max} < 1.0 \ TeV$. Potential reasons for these differences are discussed in the Appendix. A meaningful upper limit on m_H would correspond to that energy scale beyond which, if there were no H boson, predictions based on the SM Lagrangian would begin to break down. It remains to be determined just how large the deviations are corresponding to the advertised "upper limits." Here we wish to emphasize the importance for phenomenology to achieve quantitative accuracy on the implications of triviality. It will be far easier to determine experimentally whether there is a relatively weakly-coupled H ($m_H < 800 \ GeV$) than it will be to exclude a strongly coupled H.

2. STRONG YUKAWA COUPLING

In the SM, fermions get their mass via Yukawa couplings to the Higgs field. In order to account for the range of masses from the electron to the top quark, which apparently weighs more than about $50 \ GeV$, these couplings vary over 5 orders of magnitude. While technically natural,[10] unlike a light H, it remains an unattractive feature of the SM. One cannot help wondering whether there are more generations. If so, how massive might they possibly be, i.e., is there an upper limit to fermion masses? There are phenomenological constraints on the mass *splitting* between members of a doublet, such as the bottom-top splitting, coming from the magnitude of radiative corrections, e.g., to the ρ-parameter relating the vector masses $\left[\rho \equiv \frac{M_W^2}{M_Z^2 \cos^2 \Theta_W} \right]$ to the Fermi constant.[17] This line of reasoning only constrains the mass *splittings*. However, for a given H mass there appears to be an upper limit on the sum of fermion masses in order to avoid destabilizing the Higgs potential.[18] The latter result has been established only within perturbation theory and, if a Yukawa coupling becomes so large that perturbation theory breaks down, it is not clear what constraint if any applies. Indeed, very heavy fermion multiplets, as with very heavy H, suggest a strong coupling problem. In summary, while there are constraints on mass splittings within a doublet, it is not clear whether there is any phenomenological constraint on the mean mass of a multiplet.

However, if the fermion mass is large compared to the vector boson masses, then the corresponding Yukawa coupling y_f is infrared free, so it is natural to ask whether this, like the Higgs self-coupling λ, is not also trivial. This is an especially difficult question in the SM because of the interplay between the renormalization of λ and y_f, and the answer is unknown. Fermions are more difficult to treat on the lattice, especially if chiral, but because we are interested in the Higgs phase in which the fermion is described by a massive Dirac field, the underlying chiral nature of the theory may not be as important. However, including dynamical fermions on the lattice increases the already substantial demands on computing power required by lattice field theory simulations.

Below we describe a model[19] which can be more or less analytically solved via a 1/N expansion and establishes the triviality of the Yukawa coupling. This was carried out in the spirit of an existence proof and does *not* imply that the same result holds for fixed N or in the SM. This model suggests that it may not be completely crazy to begin with quenched fermions, an approximation which is of course much more amenable to simulation. Preliminary indications[20] suggest that, in the quenched approximation, the single component ϕ^4 theory with Yukawa couplings to fermions does in fact have trivial Yukawa couplings. I should mention in passing that another aspect of this question concerns anomalous gauge theories, their consistency and renormalizability,[19,21] but I will not have time to take up that application here.

The model I wish to describe has N left-handed ψ_L^l and N right-handed ψ_R^r fermions and an $N \times N$ complex Higgs field Φ_r^l. For $N = 2$, this is not the minimal model, but like a two-doublet model, the simplest extension of the minimal SM. We assume that the fermion mass of interest is large compared to the vector boson masses and that the gauge couplings can therefore be neglected compared to the Yukawa couplings. The Lagrangian we wish to consider is

$$\mathcal{L} = \bar{\psi}_L^l i\!\!\not{\partial}\psi_L^l + \bar{\psi}_R^r i\!\!\not{\partial}\psi_R^r + y\left(\bar{\psi}_L^l \Phi_L^l \psi_R^r + \text{h.c.}\right) + \mathcal{L}_H, \tag{11}$$

to which the addition of counterterms is tacitly implied. The H Lagrangian need to which the addition of counterterms is tacitly implied. The H Lagrangian need not be precisely specified but might be of the form

$$\mathcal{L}_H = Tr \partial_\mu \Phi^\dagger \partial^m u \Phi = \frac{1}{2} \left[Tr \left(\Phi^\dagger \Phi \right) - v^2 \right]^2. \tag{12}$$

Without loss of generality, we assume that $y > 0$ and that, for stability, $\lambda(\mu) > 0$ at least at large enough scales μ. There is a global $U_N \otimes U_N$ invariance spontaneously broken to a diagonal U_N with $m = yv$ the mean mass of the fermion multiplet in tree approximation. To consider variations with the fermion mass, we imagine keeping fixed the VEV $\langle \Phi \rangle \equiv v1$ as we vary y (at some arbitrary scale). We shall expand in powers of $1/N$. In order that m be of $O(1)$ in such an expansion, we hold $y^2 N$, λN, and v^2/N fixed and of $O(1)$. For this reason, it is convenient to redefine $y \to y/\sqrt{N}, \lambda \to \lambda/N, v \to v\sqrt{N}$, so that *henceforth y, λ, and v are simply independent of N*. Our goal is to determine the running coupling $y(\mu)$ to determine whether the Yukawa coupling is trivial and to study the behavior of the physical fermion mass as a function of y by calculating the pole in the fermion propagator.

We will only summarize results here; the details can be found in Ref. [19].[13] To leading order in $1/N$, there are no corrections to the proper Yukawa vertex, i.e., the only renormalization of y is due to the fermion wave-function renormalization. Thus, $\beta_y = 2y\gamma$, where γ is the anomalous dimension of the fermion field. It does in fact turn out that y is trivial and that there is a scale Λ_f at which y blows up. A feature of the $1/N$ expansion in this model is that the contribution of the Yukawa coupling to β_λ is suppressed by one power of $1/N$ and can be neglected.[14] To $O(1)$ in $1/N$, there are no radiative corrections to the renormalized fermion mass term, so that the renormalized inverse fermion propagator has the form $\mathcal{A}p - yv$ for some function A. The physical mass m is of course associated with the solution of $\mathcal{A}^2 p^2 = y^2 v^2$. Even as simple as this system is, we have only been able to solve it after further approximation. We suppose that, at scales of interest, $\lambda \ll y^2$ so

---

[13] There is a technical error there, to be treated in an Erratum, with regard to the renormalization of the scalar mass and self-coupling in which they differ from Ref. [2], but in the light Higgs approximation to be introduced below, the results of Ref. [19] are unaffected.

[14] While this makes it much simpler to analyze the model, it also is a distinct difference from a fixed N approach, where this feedback is extremely important and, as mentioned earlier, at one-loop tends to drive λ negative.

that we may ignore scalar rescattering corrections relative to Yukawa interactions.[15] Then the "wave-function renormalization" function $\mathcal{A}$ obeys the integral equation (see Fig. 3)

$$(\mathcal{A} - 1)\not{p} = y^2 \int \frac{d^4 k}{(2\pi)^4} \frac{\mathcal{A}\not{k}}{(\mathcal{A}^2 k^2 - yv)} \frac{1}{(p-k)^2} + \text{counterterm.} \qquad (13)$$

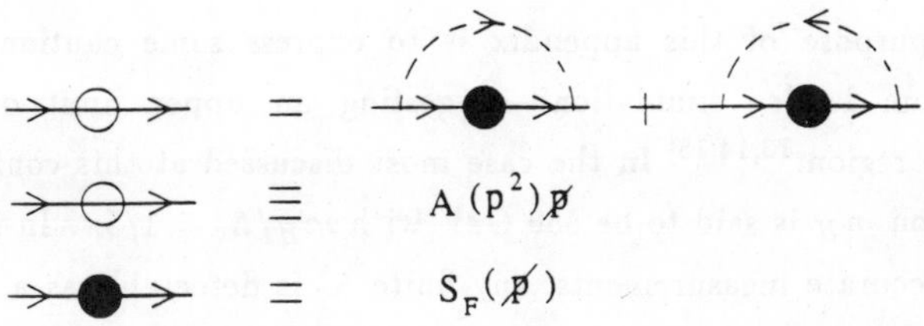

Figure 3. Diagrams depicting the integral equation that determines the propagator function $\mathcal{A}$

The specification of the counterterm determines, of course, the wave-function renormalization constant Z and thereby the anomalous dimension γ and thereby β_y. One can exploit the freedom in the choice of the finite part of the counterterm to choose

$$\mathcal{A} = \frac{y(\mu)}{Y(t; y(\mu))}, \qquad (14)$$

where Y is the running coupling constant and $t = \frac{1}{2} \ln \left(p_E^2 / \mu^2 \right)$, for Euclidean momenta $p_E^2 \equiv -p^2$. Then one may show that indeed the running coupling blows up at some finite scale $p_E^2 = \Lambda_f^2$, yet the physical fermion mass increases with increasing y [approximately as $2\pi v \ln(y)$] so that there comes a point where $m = \Lambda_f \approx 10\pi v$. If one naively assumed this were true for $N = 2$, this would correspond to an upper limit of 5.6 TeV.

---

[15] Again, this approximation is apparently only possible in an expansion in $1/N$, and it is unclear whether the fixed N case will share the features which emerge in this model.

82

Acknowledgements

I wish to acknowledge the participation of D.N. Williams in much of the work discussed here. I would like to thank P. Hasenfratz, J. Kuti, L. Lin, M. Lüscher, H. Neuberger, Y. Shen, and P. Weisz for extremely helpful comments and discussions. Needless to say, our views might well differ on certain points. This work was supported in part by the U. S. Department of Energy.

APPENDIX

The purpose of this appendix is to express some caution about results obtained from lattice simulations suggesting an upper limit on m_H in the perturbative region.[13,14,15] In the case most discussed at this conference, [13] the upper limit on m_H is said to be 550 GeV with $m_H/\Lambda_c = 1/5\pi$. In principle, given arbitrarily accurate measurements, any finite Λ_c is detectable as a deviation from perturbative predictions based on the SM Lagrangian. Therefore, the meaning of any upper limit is a quantitative issue depending on (1) the accuracy of experimental tests and (2) the physics underlying Λ_c.

(1) With regard to the first issue, the difficulties observing the interaction of longitudinal vector bosons (née Goldstone bosons), e.g., W^+W^- scattering, even with the design parameters of the SSC, may make it difficult to identify deviations from the SM unless they are rather large (say, 50%). As a second example, consider calculations of the usual kinds of radiative corrections to the ρ-parameter relating the Fermi constant G_F to the vector boson masses frequently discussed as a precision test of the Standard Model.[17] Typically, the dependence on m_H or on fermion mass splittings (such as the difference between the top and bottom quark masses) are at the level of less than 1% even when these parameters are so large as to reach the limits of perturbation theory. Precision measurements apparently will not exceed an accuracy of 0.1%, even for the vector masses themselves.

(2) With regard to the second issue, it may very well be that if the lattice were not a calculational artifact but rather a true representation of the physics beyond Λ_c, deviations from the SM would become apparent as advertised.[13] However, if the physics underlying the SM remains, for example, Lorentz invariant, then one would expect corrections to SM predictions to be of order $(m_H/\Lambda_c)^2$ which, for the value quoted above, would correspond to $(1/5\pi)^2 = 0.4\%$

for Higgs interactions [and much smaller in measurable radiative corrections proportional to $\alpha \ln(m_H)$].[16]

A physically popular constituent model, technicolor,[23] suggests further reason for caution. There it would appear that $\Lambda_c = \Lambda_{TC} \approx 1 TeV$ and, in the simplest version of this model, there is no H-boson below this scale. How near to Λ_{TC} the O_4 model accurately describes interactions isn't known, but there certainly is not a perturbative upper limit on the H.

Three other reasons for concern are these: First, despite what we have heard about the method of extracting m_H for near zero momenta, I believe a better job needs to be done concerning finite size effects and divergences due to nearly massless Goldstone bosons.[24] Secondly, the representation of the continuum on lattices as small as those currently being used needs further investigation; in particular, do the several resonances below m_H accurately reproduce the H decay width?[17] Thirdly, in another talk at this Workshop, it has been shown that there is substantial variation in the results depending on the lattice type.[25]

At least until the several issues are clarified, we should be cautious about taking current numerical results too seriously.

REFERENCES

1. J.M. Cornwall, D.N. Levin, and G. Tiktopoulos, Phys. Rev., D10, (1974) 1145.
   B.W. Lee, C. Quigg, and H.B. Thacker, Phys. Rev., D16, (1977) 1519.
   M. Chanowitz and M.K. Gaillard, Nucl. Phys., B261, (1985) 379.

2. M.B. Einhorn, Nucl. Phys. B, 246B, (1984) 75 a more pedagogical discussion may be found in M.B. Einhorn, 1984 TASI Lectures in Elementary Particle Physics, ed. D. Williams, Ann Arbor, 1984.

3. G. 't Hooft, Nucl. Phys., B72, (1974) 461.
   G. 't Hooft, Comm. Math. Phys., 86, (1982) 449.
   S. Coleman, Int. School of Subnuclear Physics, Erice.
   E. Witten, Nucl. Phys., B160, (1979) 57.

4. For reviews, see S.K. Ma, Rev. Mod. Phys., 45, (1973) 589.
   See lectures in, Phase Transitions and Critical Phenomena, Vol. 6, eds. C. Domb and M.S. Green (New York: Academic Press, 1986).

---

[16] This echoes concerns previously expressed by H. Neuberger.[22]

[17] I appreciate comments from M. Lüscher and P. Weisz for emphasizing this point to me. See U. Wiese, in preparation.

5. E. Brézin, U.C. le Guillou, and J. Zinn-Justin, Phase Transitions and Critical Phenomena, Vol. 6, eds. C. Domb and M.S. Green (New York: Academic Press, 1986).

6. A more recent discussion reviewing problems with earlier attempts at a solution is W. Bardeen and M. Moshe, Phys. Rev., D28, (1983) 1372.

7. B.W. Lee, Chiral Dynamics (Gordon & Breach, 1972).

8. R. Casalbuoni, D. Dominici, and R. Gatto, Phys. Lett., 147B, (1977) 419.

9. M.B.Einhorn and D.N. Williams, On the Lattice Approach to the Maximum Higgs Boson Mass, UM-TH-88-07, May 1988.

10. G. 't Hooft, Lecture II, in: Recent Developments in Gauge Theories, eds. G. 't Hooft, et al., (Plenum, N.Y., 1980) Proc. NATO ASI, Cargèse.

11. B.W. Lee, C. Quigg, and H.B. Thacker, Phys. Rev. Lett., 38, (1977) 883.
    B.W. Lee, C. Quigg, and H.B. Thacker, Phys. Rev., D16, (1977) 1519.

12. J. Bjorken and S.D. Drell, Relativistic Quantum Fields (N.Y.: McGraw-Hill, 1965) 139.

13. J. Kuti, L. Lin, and Y. Shen, UCSD/PTH 87-18 & 87-19 (November, 1987) and contributions to this Workshop.

14. H.G. Evertz et al., Nucl. Phys., B285 [FS19], (1987) 590.
    A. Hasenfratz et al., Phys. Lett., B199, (1987) 531.

15. A semi-analytic approach based on the high temperature expansion on the lattice yields similar conclusions for the one-component model. See M. Lüscher and P. Weisz, Nucl. Phys., B290 [FS20], (1987) 25.
    M. Lüscher and P. Weisz, Nucl. Phys., B290 [FS20], (1987) 65.
    M. Lüscher, NATO Advanced Study Institute on Non-Perturbative Quantum Field Theory Cargese, 1987, DESY 87-159, Dec. 1987.

16. An analytic calculation for O_4 based on an approximate renormalization group transformation provides additional supporting evidence. See P. Hasenfratz and J. Nager, Z. Physik, C37, (1988) 477.

17. Apparently, the mass of the top must be less than about 200 GeV. For a recent review of data, See U. Amaldi et al., Phys. Rev., D36, (1987) 1385.

18. H.D. Politzer and S. Wolfram, Phys. Lett., 82B, (1979) 242.
    H.D. Politzer and S. Wolfram, Phys. Lett., 83B, (1979) 421.
    P.Q. Hung, Phys. Rev. Lett., 42, (1979) 873.
    M.J. Duncan, R. Philippe, and M. Sher, Phys. Lett, 153B, (1985) 165.
    M. Lindner, Zeit. Phys., C31, (1986) 295.

19. M.B. Einhorn and G.J. Goldberg, Phys. Rev. Lett., 57, (1986) 2115.

20. J. Shigemitsu, Phys. Lett., 189B, (1987) 164.

21. E.D'Hoker and E. Farhi, Nucl. Phys., B248, (1984) 59.
    K. Harada and I. Tsutsui, Phys. Lett., B183, (1987) 311.

K. Harada and I. Tsutsui, Prog. Theor. Phys., 78, (1987) 675.

L.D. Faddeev, Phys. Lett., B145, (1984) 81.

L.D. Faddeev and S.L. Shatashvili, Phys. Lett., 167B, (1986) 225.

22. H. Neuberger, Phys. Lett., 199B, (1987) 536.

23. For reviews of this subject, see K.D. Lane and M.E. Peskin, lectures given at 15th Rencontre de Moriond, published in Moriond Conf. 1980 v. 2, p. 469.

E. Farhi and L. Sussking, Phys. Rept., 74, (1981) 277.

G.L. Kane, lectures given at Les Houches Summer School in Theoretical Physics, Les Houches, France, 1981.

R.K. Kaul, Rev. Mod. Phys., 55, (1983) 449.

24. H. Neuberger and U.M. Heller, See the contributions to these Proceedings.

25. G. Bhanot and K. Bitar, SCRI Preprint FSU-SCRI-88-50 and these Proceedings.

GAUGE GLASSES - SYMMETRY BREAKING WITHOUT HIGGS?

Simon Hands

Department of Theoretical Physics
1 Keble Road
Oxford, OX1 3NP
United Kingdom

ABSTRACT

Gauge glasses are described as generalizations of spin glass models, and their potential property of gauge symmetry breakdown at weak coupling is discussed. Some simple calculations on abelian models are presented.

Gauge glasses are lattice gauge theories with quenched random couplings located on each plaquette; as such they are geometrical extensions of the spin glass models long studied in condensed matter physics. The models were first suggested by H.B. Nielsen and collaborators[1] as an idealization of a gauge theory vacuum which is severely disordered at some fundamental scale, presumably the Planck length. In this case the lattice spacing need no longer be thought of as an ultraviolet cutoff which must ultimately be removed from the calculation of physical quantities.

To begin with, consider the Hamiltonian for a typical model of a spin glass:

$$H = - \sum_{<ij>} J_{ij}\sigma_i\sigma_j. \qquad (1)$$

Here σ_i are dynamical spin variables defined on the sites of a regular lattice, J_{ij} is a quenched random coupling drawn from some distribution, (the most frequently considered are either a normal one centered on zero parameterized by its width, or a bimodal '$\pm J$' one, i.e., the bonds have constant magnitude but random sign), and the sum over $< ij >$ is limited to nearest neighbors. J_{ij} is quenched in the sense that the timescale over which the couplings vary is much greater than that of the σ_i.

The width of the coupling distribution is inversely proportional to the temperature of the system. An essential feature is the presence of competing interactions, since J_{ij} can be positive or negative.

An important symmetry of H is gauge invariance,[2] namely H is preserved under

$$\sigma_i \mapsto T_i \sigma_i; \quad J_{ij} \mapsto T_i J_{ij} T_j^{-1}, \tag{2}$$

where T_i is some suitable transformation (e.g., for Ising spins, $T_i \epsilon \{+1, -1\}$). Gauge invariance is important because in the thermodynamic limit many measurable quantities defined as expectation values over an ensemble of configurations σ_i lose any dependence on the detailed spatial distribution of the J_{ij}. In a finite system, therefore, measurements must be averaged over many bond distributions, and under this procedure only gauge invariant quantities should have meaning. The simplest gauge invariant function of the random bond strengths is the frustration Φ_{ijkl} defined on plaquettes of the lattice by

$$\Phi_{ijkl} = \prod_{<ij> \in ijkl} \text{sign}(J_{ij}). \tag{3}$$

Suppose, in an Ising spin glass with bimodal bond distribution, that $\Phi_{ijkl} = -1$ on a particular plaquette. It is impossible to arrange spins at the corners so that all four bonds are simultaneously satisfied: one bond is always 'frustrated'. In this case, therefore, the number of ground states of the isolated plaquette is increased from two to eight. This is a general feature of the spin glass problem: gauge invariant or 'serious' disorder leads to many distinct ground states. In general, the energy surface in the space of all possible states is believed to have a complicated topography with an exponentially increasing number of deep valleys separated by large energy barriers. The behavior of spin glasses is characterized by the system settling into one of these deep minima at sufficiently low temperatures and becoming trapped and unable to exhibit ergodic behavior. The existence and description of this spin glass transition has been a major focus of study (see [3] and references therein).

The gauge glass can be defined on a d dimensional hypercubic euclidean lattice by the following action;

$$S = \sum_{<ijkl>} \text{Re} \left[\beta_{ijkl} \text{Tr}(U_{ij} U_{jk} U_{kl} U_{li}) \right]. \tag{4}$$

where as usual the links U_{ij} are elements of some gauge group and the quenched random couplings β_{ijkl} are defined on plaquettes and are in general drawn from a distribution of complex numbers. As well as the usual explicit local gauge invariance, S has another symmetry in analogy with what we called 'gauge invariance' in the spin glass case, namely

$$U_{ij} \mapsto \tau_{ij} U_{ij}; \ \beta_{ijkl} \mapsto \tau_{ij}^{-1} \tau_{jk}^{-1} \tau_{kl}^{-1} \tau_{li}^{-1} \beta_{ijkl}, \tag{5}$$

where the transformation τ_{ij} lies in the center of the gauge group. Once again we may define a (complex) frustration variable Φ_{cube} sited on elementary 3-cubes which is an invariant of this 'extended gauge symmetry':

$$\Phi_{\text{cube}} = \prod_{<ijkl> \in \text{cube}} \frac{\beta_{ijkl}}{|\beta_{ijkl}|}. \tag{6}$$

For models with a distribution of β with fixed modulus, it is straightforward to see that once again the system may have many ground states if some $\Phi_{\text{cube}} \neq 1$. This is because any configuration of U_{ij} must satisfy a Bianchi identity[4]

$$\prod_{<ijkl> \in \text{cube}} U_{ijkl} = 1, \tag{7}$$

where

$$U_{ijkl} = U_{ij} U_{jk} U_{kl} U_{li}.$$

For instance, if the gauge group is Z_2, then if $\Phi_{\text{cube}} = -1$, assuming all plaquettes were independent the action-optimal configuration would have an odd number of plaquettes in the cube equal to -1, in contradiction to (7). Thus at least one face of the cube is frustrated and, just as before, this leads to a multiplicity of ground states, and the possibility of a gauge glass phase transition where the system becomes locked into a particular local energy well.

Nielsen and co-workers[4] have argued that in the case where the gauge group G is large, continuous, and non-abelian, the resulting ground state will in general not be invariant under gauge transformations of the type

$$T_i U_{ij} U_{jk} U_{kl} U_{li} T_i^{-1}$$
$$T_i \epsilon G'. \tag{8}$$

This constrains gauge symmetries G' which do survive the continuum limit, to participate in physics at energies much smaller than the Planck scale, to be subgroups of G having centers which are continuous and connected. As the width of the $|\beta|$ distribution grows large (we can refer to this as the weak coupling limit in analogy with ordinary lattice gauge theory), the gauge glass relaxes into its preferred vacuum state, and in this case deviations of each U_{ijkl} away from the value picked out by locally optimising S on each plaquette can be accommodated within the center of G', and so the symmetry of equation (8) holds, with the Us considered as elements of G'. The gauge glass may thus contain a natural mechanism for spontaneous gauge symmetry breaking $G \rightarrow G'$ without matter fields. Moreover, there are group-theoretic arguments[5] to suggest that the most likely survivor of such a breakdown would be the standard model group $S(U(2)\otimes U(3))$[6] plus other groups which would confine and hence remain unobserved at low energies.

Clearly, in spite of symmetry arguments, the existence of a gauge glass phase in the limit of large $|\beta|$ is also a dynamical question. One can begin to address it in the context of simple abelian models using techniques developed in spin glass physics. Consider a Z_2 gauge glass model in which the β_{ijkl} are normally distributed about zero with width K. One widely used analytical approach is that of replica mean field theory; namely, the disorder in the system is averaged over, by first evaluating the partition function Z^n for a system of n non-interacting replica fields U_{ij}^α on each link, and then defining the quenched free energy by the limit

$$\bar{F} = \overline{\ln Z} = \lim_{n \to 0} \frac{\bar{Z}^n - 1}{n}. \tag{9}$$

$\bar{Z}^n$ is easily found from (4) by a gaussian integration over the β_{ijkl}:

$$\bar{Z}^n \propto \mathrm{Tr}\exp K^2 \sum_{<ijkl>} \sum_{\alpha<\beta=1,n} U_{ij}^\alpha U_{jk}^\alpha U_{kl}^\alpha U_{li}^\alpha U_{ij}^\beta U_{jk}^\beta U_{kl}^\beta U_{li}^\beta. \tag{10}$$

The non-trivial nature of the quenched average is indicated by the generation of interactions between replicas. The mean field solution of this model follows lines similar to that found by Gross and Mézard in their treatment of the random energy model[7]. A phase transition is found at a critical coupling width $K_c = 2(\ln 2/d)^{\frac{1}{2}}$: the different phases are characterized by an overlap function

$$q^{\alpha\beta} \equiv \frac{1}{Nd} \sum_{<ij>} U_{ij}^\alpha U_{ij}^\beta, \tag{11}$$

where N is the number of lattice sites. In the small K phase, $q^{\alpha\beta} = 0$, and the free energy is identical to that of the unfrustrated Z_2 lattice gauge theory in its strong coupling phase. For $K > K_c$, replica symmetry breaking occurs, and $q^{\alpha\beta}$ assumes values 0 and ~ 1, indicating large self-overlap of replicas or, equivalently, that the system has almost completely frozen into a low energy pure state. This is the gauge glass phase. Unfortunately, more careful calculation[8] reveals that this solution is only stable as $n \to 0$ for $d > 5$. This prediction for the lower critical dimension is backed up by a simple real space renormalisation group treatment, in which distributions of β_{ijkl} are evolved under blocking using a Migdal-Kadanoff procedure. The possible fixed points can be characterized by distribution width K^* equal to 0 or ∞, corresponding to the strong coupling or gauge glass phases. The second type of fixed point is only observed for $d \geq 6$ in a numerical study. The form of the fixed point coupling distribution does not noticeably deviate from a gaussian. Further details are presented elsewhere[8].

Despite the disappointing implication that gauge symmetry breaking may only be a possibility in dimensionalities greater than four, we can go on to build up a qualitative picture of serious disorder in the context of the simplest continuous gauge symmetry, $U(1)$, and work in a manifestly extended gauge invariant formalism. For a model with couplings of fixed modulus K but random arguments θ_{ijkl}, the partition function is

$$Z_K\{\Phi_{\text{cube}}\} = \int_0^{2\pi} \prod_{<ij>} d\omega_{ij} \exp\left(K \sum_{<ijkl>} \cos\Big(\sum_{<ij>\in ijkl} \omega_{ij} + \theta_{ijkl}\Big)\right), \quad (12)$$

with

$$\Phi_{\text{cube}} \equiv \Phi_{\mu\nu\lambda i} \equiv \sum_{<ijkl>\in\text{cube}} \theta_{ijkl} \,(\text{mod}2\pi).$$

The greek letters denote lattice direction. We can follow the standard procedure[9,10] of using the Villain approximation, integrating out the link variables ω_{ij}, expressing the resulting constrained sum in terms of integer valued variables $n_{\mu i}$ defined on the links of the dual lattice, and then resumming using the Poisson formula. The result in four dimensions is

$$Z\{\Phi_{\mu i}\} = Z_{sw} \sum_{\{m\}} \exp\left(-2\pi^{2K} \sum_{ij} (\Phi + m)_{\mu i} v_{\mu\nu}(\bar{i} - \bar{j})(\Phi + m)_{\nu j}\right), \quad (13)$$

where

$$\Phi_{\mu\tilde{i}} \equiv \frac{1}{2\pi}\varepsilon_{\mu\nu\lambda\kappa}\Phi_{\nu\lambda\kappa i}.$$

The partition function is written in terms of integer valued excitations $m_{\mu\tilde{i}}$ (which may be interpreted as world lines of magnetic monopoles in four dimensions), and the fractionally charged quenched frustration flux $\Phi_{\mu\tilde{i}}$, both defined as vectorlike variables on the dual links, interacting via a coulomb interaction. The (gauge-fixed) coulomb propagator $v_{\mu\nu}$ has the asymptotic behavior $v(\tilde{i} - \tilde{j}) \sim |\tilde{i} - \tilde{j}|^{2-d}$ in d dimensions. The original gauge invariance of the model imposes an important additional constraint:

$$\Delta_\mu(\Phi + m)_{\mu\tilde{i}} = 0. \tag{14}$$

Thus the vacuum consists of quasi-continuous trees of frustration flux which both seed monopole excitations (in the unfrustrated theory the monopoles are constrained by (14) to form closed loops) and screen monopole-monopole interactions, thus reducing the activation energy for loops to form. One might expect a dense system of frustrations to suppress the Kosterlitz-Thouless type mechanism for the phase transition of unfrustrated U(1) lattice gauge theory in four dimensions,[10] and possibly drive the lower critical dimension upwards from four.

Z_{sw} in equation (13) is the partition function for spin waves, and is given in terms of a continuous link degree of freedom $\phi_{\mu\tilde{i}}$, arising from the Poisson resummation:

$$Z_{sw} = \int_{-\infty}^{+\infty} \prod_{\mu i} d\phi_{\mu\tilde{i}} \exp\left(-\frac{1}{4K}\sum_i [\Delta_\lambda\phi_{\kappa\tilde{i}} - \Delta_\kappa\phi_{\lambda\tilde{i}}]^{2 + S_{\text{gauge fixing}}}\right). \tag{15}$$

It has no role in describing the phase structure of the model, but describes massless excitations of the gauge degrees of freedom, namely photons (in the weak coupling limit of U(1) lattice gauge theory an identical term supposedly reproduces free non-compact QED in the continuum limit). Thus in this trivial example we see how the gauge glass does contain a perfectly good continuum gauge theory, albeit completely separated from any measurable spatial correlation function, and also without any gauge symmetry breaking.

To conclude, gauge glasses are an interesting generalisation of spin glasses, and there is ample room for improvement on the techniques discussed here in examining their behavior at weak couplings. For example, the mean field formalism, powerful as it is, is couched in terms of a gauge variant order parameter: clearly this is not suitable to investigate the symmetry properties of a glassy phase of a non-abelian theory. It appears difficult to reconcile gauge symmetry collapse as a consequence of random dynamics with the simple models presented here, in four dimensions. However, we should remember that the symmetry breaking mechanism proposed is essentially non-abelian, and hence beyond the scope of the methods we have used.

REFERENCES

1. D.L. Bennett, N. Brene and H.B. Nielsen, Random Dynamics, Talk given at 2nd Nobel Symposium on Elementary Particle Physics, Marstrand.

2. G. Toulouse, Commun. Phys. 2, (1977) 115.

3. K. Binder and A.P. Young, Rev. Mod. Phys. 58, (1986) 801.

4. H.B. Nielsen and D.L. Bennett, The Gauge Glass: A Short Review, NORDITA preprint 85/23.

5. H.B. Nielsen and N. Brene, Nucl. Phys. B224, (1983) 396.

6. L. O'Raifeartaigh, Group Structure of Gauge Theories Cambridge University Press, U.K., (1986).

7. D.J. Gross and M. Mézard, Nucl. Phys. B240 [FS12], (1984) 431.

8. S.J. Hands, Abelian Gauge Glasses, Oxford preprint 91/87, December, 1987 To appear in Nucl. Phys..

9. J.V. José, L.P. Kadanoff, S. Kirkpatrick and D.R. Nelson, Phys. Rev. B16, (1977) 1217.

10. T. Banks, J. Kogut and R. Myerson, Nucl. Phys. B129, (1977) 493.

THE EFFECT OF STRONG YUKAWA COUPLING ON THE SCALAR-FERMION MODEL

Anna Hasenfratz

Supercomputer Computations Research Institute
The Florida State University
Tallahassee, FL 32306-4052

ABSTRACT

The one component scalar φ^4 model coupled to flavor singlet fermions are studied in the quenched approximation. In the broken phase of the scalar field we found an unexpected (first order) phase transition in the Yukawa coupling.

1. INTRODUCTION

In this talk I report on a work done recently in collaboration with Thomas Neuhaus.[1] We studied the 1-component scalar model coupled to flavor singlet fermions via Yukawa interaction.

As a consequence of the triviality of the scalar φ^4 model[2] the scalar sector of the WS model is a trivial free field theory in the infinite cut-off limit.[3] However as an effective model with a large but finite cut-off it can describe an interacting theory. In the effective model the mass of the scalar particle increases with decreasing cutoff, therefore from the requirement that $m_s < \Lambda_{cut}$ in a sensible effective theory one gets an upper bound for the scalar particle mass.[4] The scalar sector of the model is prather well understood by now, analytical and numerical results are supporting the above picture and predicting an upper bound for the Higgs mass $m_H = 600 GeV$ at $\Lambda_{cut} = 3\pi m_H$.[4-7]

The effect of fermions is less understood however. There are studies concerning chiral symmetry breaking in gauge-Higgs models[8] and lately some works have been reported concerning the effect of the strong Yukawa coupling.[9-12] Why is the strong Yukawa coupling region interesting? In the scalar-fermion model the Yukawa coupling (y) is non-asymptotically free, its β function in the perturbative (small y) region behaves the same way as the quartic scalar coupling. Adding the gauge field can drastically change this behaviour. For example, introducing the SU(3) strong interaction with at least 2 generations of fermions will turn the Yukawa coupling asymptotically free. In addition perturbation theory predicts the existence of an infrared fixed point in this case. Nevertheless it was generally assumed that the strong Yukawa coupling region is determined by the scalar sector only. If so, and the Yukawa coupling behaves similarly to the scalar self coupling even for strong y (i.e. the β function is positive for all values of the Yukawa coupling in the scalar-fermion model) there is an upper bound for the fermion mass in an effective model in the same way as there is an upper bound for the scalar mass.[10]

We have studied the model with the action

$$
\begin{aligned}
S =& S_{sc} + \sum_{n,\mu} \overline{\psi}_n \frac{\gamma\mu}{2}(\psi_{n+\mu} - \psi_{n-\mu}) + y \sum_n \overline{\psi}_n \psi_n \varphi_n \\
S_{sc} =& - \kappa \sum_{n,\mu} \varphi_n(\varphi_{n+\mu} + \varphi_{n-\mu}) + \lambda \sum_n (\varphi_n^2 - 1)^2 + \sum_n \varphi_n^2
\end{aligned}
\tag{1}
$$

We found that the expectations based on the perturbative expansion do not hold. While we reproduced the perturbative result in the weak Yukawa coupling region, for strong coupling the renormalized Yukawa coupling (or the physical fermion mass) is not running to zero with increasing cutoff but increases itself. The two phases are separated (at least in the quenched approximation) by a strong first order phase transition.

2. THE BETA FUNCTION AT STRONG AND WEAK YUKAWA COUPLING; THE QUENCHED ACTION

In the perturbative region the Yukawa coupling is non-asymptotically free, its β function is positive, $\beta(y_r) \equiv dy_r/dt \sim y_r^3$. As the physical fermion mass is $m_f = y_r \langle\varphi\rangle_r$ at tree level, the function $\beta(m_f) \equiv dm_f/dt \sim m_f^3$ is positive for small y as well. When the cut-off is taken to infinity m_f approaches zero.

A simple lowest order strong coupling expansion shows however that for large y the situation is different. Consider the fermion propagator $< \psi_{x,0}^1 \overline{\psi}_{x,t}^1 >$. Expanding the action in the hopping term at lowest order in $1/y$ we obtain

$$\left\langle \psi_{x,0}^1 \overline{\psi}_{x,t}^1 \right\rangle = \frac{1}{Z} \int D(\overline{\psi}_i \psi_i \varphi_i) \cdot \psi_{x,0} \prod_{i=0}^{t} \left(\overline{\psi}_{x,i}^1 \frac{\gamma_0}{2} \psi_{x,i+1} \right)^{1,1} \overline{\psi}_{x,t}^1 e^{-S_{sc} - y \sum_n \overline{\psi}_n \psi_n \varphi_n}$$

$$= \frac{1}{Z'} \int D\varphi_i \left(\prod_V \varphi_n^4 \right) \left(\prod_{i=0}^{t} \frac{1}{2y\varphi_i} \right) e^{-S_{sc}}$$

(2)

where*

$$Z' = \int D\varphi_i \left(\prod_V \varphi_n^4 \right) e^{-S_{sc}} = \int D\varphi_i e^{S_{sc} + \sum_n \ln \varphi_n^4}$$

(3)

The expectation value

$$\left\langle \prod_{i=0}^{t} \frac{1}{\varphi_i} \right\rangle' = \frac{1}{Z'} \int D\varphi_n \left(\prod_{i=0}^{t} \frac{1}{\varphi_i} \right) e^{-S_{sc} + \sum_n \ln \varphi_n^4}$$

(4)

has to be computed numerically. MC simulations show that $\Phi = \lim_{t\to\infty} \left\langle \prod_0^t \frac{1}{\varphi_i} \right\rangle^{-1/t}$ is finite, $\Phi > \langle \varphi \rangle$ and increases as the scalar correlation length increases. That means that the physical fermion mass in the strong coupling region $m_f = y\Phi$ is larger than the bare and increases as the cutoff is removed to infinity. The function $\beta(m_f)$ is negative. It is obvious that $\beta(m_f)$ has a zero or is singular at some finite value of the Yukawa coupling separating the two phases of the system.

The fact that the expectation value Φ in (4) is computed with a modified scalar action leads us to an important observation. We are working in the quenched approximation, i.e. in the functional integral we neglect the effect of closed fermion loops. However it cannot be implemented in the ususal way simply by generating configurations with the original scalar action and calculating scalar expectation values. The expectation value of the fermion propagator which is a specific matrix element of the inverz fermion matrix is singular at $\langle \varphi \rangle = 0$ the same way as Eq(2) is singular if we neglect the term $\prod_V \varphi_n^4$. In other words it means that approximating the fermion determinant by 1 is unacceptable in this case. Instead we approximate it by $\prod_V y\varphi_n^4$. This way, although we still neglect the closed fermion loops, the functional integral is well defined. In calculating expectation values the y dependence cancels from the new term and we end up with a modified scalar action

---

* The exponent '4' in $\left(\prod_V \varphi_i^4 \right)$ reflects the fact that we are working with 4-component Dirac fermions.

$$S_{\text{mod}} = S_{sc} - \sum_n \ln \varphi_n^4 \tag{5}$$

Although we are working in the quenched approximation, as our scalar action is modified, the scalar observables like $m_s / \langle \varphi \rangle$ will be changed. However, it won't change the upper bound prediction for the Higgs mass as it is obtained from the $\lambda = \infty$ limit where the modified action is identical to the original one.

3. MONTE-CARLO SIMULATIONS AND RESULTS

We simulated the (modified) scalar action at $\lambda = \infty, 5.0$ and 2.0 on $8^3 \times 16$ lattices at several κ values corresponding to a scalar field expectation value $< \varphi > = 0.3 - 0.8$. At each coupling we performed typically 100 to 200 fermion matrix inversion on scalar configurations separated by 1000-2000 MC updates. Here we report only the results obtained using naive fermions.

We calculated zero momentum fermion propagators by summing over time slices both at the location of ψ and $\overline{\psi}$. This decreased our statistical error significantly but as a consequence we could only calculate the time-slice averaged version of $< \overline{\psi}\psi >$

$$\Psi = \frac{1}{N_s^6} \left\langle \sum_x \overline{\psi}_{x,0} \sum_x \psi_{x,0} \right\rangle \tag{6}$$

This quantity is close to $< \overline{\psi}\psi >$ in the large y region but differs significantly in the perturbative regime.

In all investigated cases we found strong signal for a first order phase transition in the region $y = 1.2 - 1.8$ varying surprisingly little with the parameters of the scalar model. In the following we discuss in more detail the $\lambda = \infty$ case and also show the results for finite λ.

Both the fermion mass and the quantity Ψ shows strong discontinuity at $y \sim 1.5$ (Figure 1. and 2.). In the weak coupling ($y < 1.2$) region we find agreement with the perturbative predictions. The fermion mass decreases as the correlation length increases (i.e. the cut-off is going to infinity) as it is expected from a non-asymptotically free quantity. Figure 3 shows $1/y_r^2$ as the function of $1/y_b^2$, y_r defined as $y_r = m_f / \langle \varphi \rangle_r = m_f Z^{1/2} / \langle \varphi \rangle$. 1-loop perturbation theory predicts $1/y_r^2 = 1/y_b^2 + const$. Figure 3 shows a linear dependence of $1/y_r^2$ on $1/y_b^2$ but with a slope slightly larger than 1. The reason of this small disagreement is

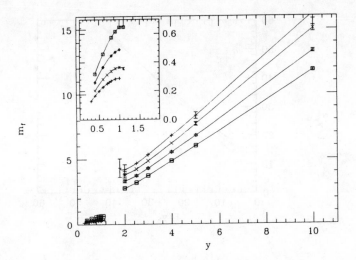

Figure 1. The fermion mass as the function of the bare Yukawa coupling at $\lambda = \infty$. The different symbols correspond to different scalar couplings: vertical cross:$\kappa = 0.078(m_s = 0.66(2))$, diagonal cross:$\kappa = 0.080(m_s = 0.91(3))$, diamond:$\kappa = 0.085(m_s = 1.38(6))$, square:$\kappa = 0.095,(m_s = 2.1(2))$. The insert enlarges the small Yukawa coupling region.

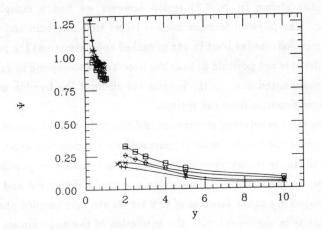

Figure 2. Ψ as the function of the bare Yukawa coupling at $\lambda = \infty$. The symbol notation is the same as in Figure 1).

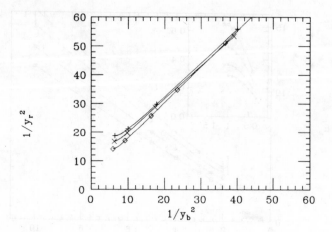

Figure 3. $1/y_r^2$ as the function of $1/y_b^2$ for weak bare Yukawa couplings at $\lambda = \infty$. The symbol notation is the same as in Figure 1).

probably that in calculating y_r we substituted $Z = 1.0$ while analytic results and high statistics MC data gives $Z \sim 0.9$ in the investigated region.[5]

In the strong ($y > 1.8$) region however we find a completely different behaviour. The physical fermion mass is larger than the bare and increases with increasing cut-off. As (at least in our quenched approximation) the phase transition is first order it is not possible to tune the bare Yukawa coupling to get a finite value for the renormalized one. In the infinite cut-off limit the fermion mass is infinite, the fermions decouple from the system.

The mass-generation mechanism is different in the two phases . In the weak coupling phase the fermion mass is generated via the non-zero expectation value of $< \varphi >$, while in the strong y phase the $< \overline{\psi}\psi >$ condensate is responsible for the mass generation. In the strong coupling region one would expect $\overline{\psi}\psi$ and Ψ to be close. Figure 4 shows m_f as the function of $y^2\Psi$ for the strongly coupled phase. The linear dependence is in agreement with the prediction of the approximate calculation of Ref [1] suggesting that the mass generation mechanism is indeed due to the vacuum condensate in this phase.

Figure 5. and 6. are the same as Figure 1. for $\lambda = 5.0$ and 2.0. We find the

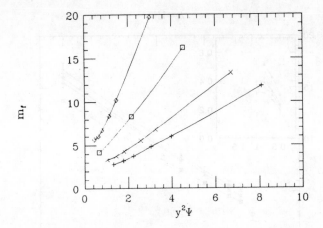

Figure 4. The fermion mass as the function of $y^2\Psi$ for large bare Yukawa couplings at $\lambda = \infty$. The symbol notation is the same as in Figure 1).

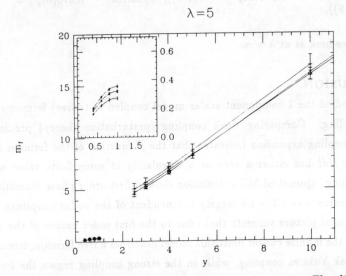

Figure 5. The same as Figure 1) but at $\lambda = 5.0$. The different symbols correspond to different scalar couplings: vertical cross:$\kappa = 0.073(m_s = 0.61(3))$, diagonal cross:$\kappa = 0.074(m_s = 0.77(3))$, diamond:$\kappa = 0.075(m_s = 0.90(3))$.

100

$\lambda = 2$

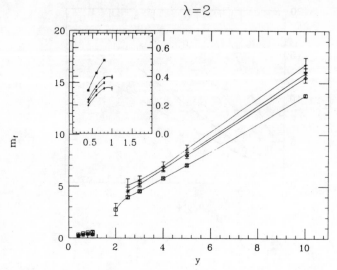

Figure 6. The same as Figure 1) but at $\lambda = 2.0$. The different symbols correspond to different scalar couplings: vertical cross:$\kappa = 0.068(m_s = 0.79(2))$, diagonal cross:$\kappa = 0.069(m_s = 0.91(3))$, diamond:$\kappa = 0.070(m_s = 1.00(5))$, square:$\kappa = 0.075(m_s = 1.52(9))$.

same structure here as at $\lambda = \infty$.

4. CONCLUSION

We studied the 1-component scalar model coupled to (naive) fermions with Yukawa coupling. Comparing weak coupling (perturbation theory) predictions and strong coupling expansion indicates that the β function of the fermion mass, $\beta(m_f) = dm_f/dt$ has either a zero or a singularity at some finite value of the Yukawa coupling. Quenched MC calculation showed first order phase transition at the coupling value $y \sim 1.2 - 1.8$ largely independent of the scalar couplings. The emerging physical picture suggests that (due to the first order nature of the phase transition) in the infinite cut-off limit this model describes a free fermion, free scalar model for weak Yukawa coupling, while in the strong coupling region the fermion mass gets infinite, the fermions decouple and the model describes a model with free scalars only. Although it is possibile that in the unquenched, full model there is no first order phase transition in the Yukawa coupling, there will be 2 separate phases even in that case. A phase corresponding to the perturbative regime where the fermion mass decreases with increasing cut-off and an other phase for strong bare Yukawa coupling where the fermion mass increases with increasing cut-off.

Acknowledgements

This work was supported by the Florida State University Supercomputer Computations Research Institute which is partially funded by the U.S. Department of Energy through Contract No. DE-FC05-85ER250000, and the Florida State University Computing Center for allocation of computer time on the ETA[10].

References

1) A. Hasenfratz and T. Neuhaus, preprint FSU-SCRI-88-55

2) K.G. Wilson, Phys. Rev. B4 (1971) 3184.
   K.G. Wilson and J. Kogut, Phys. Rep., 12C (1974) 76.
   M. Aizenman, Phys. Rev. Lett., 47 (1981) 1; Commun. Math. Phys. 86 (1982) 1.
   G.A. Baker and J.M. Kincaid, J. Stat. Phys. 24 (1981) 469.
   D.C.. Brydges, J. Fröhlich and T. Spencer, Commun. Math. Phys., 83 (1982) 123.
   J. Fröhlich, Nucl. Phys. B200 (1982) 281.
   A.D. Sokal, Ann. Inst. H. Poincare A37 (1982) 317.
   M. Aizenman and R. Graham, Nucl. Phys. B225 [FS9] (1983) 261.
   C. Aragaō de Carvalho, S. Caracciolo and J. Fröhlich, Nucl. Phys. B215 (1983) 209.
   K. Gawedzki and A. Kupiainen, Phys. Rev. Lett. 54B (1985) 92.
   B. Freedmann, P. Smolensky and D. Weingarten, Phys. Lett 113B (1982) 481.
   D.J.E. Callaway and R. Petronzio, Nucl. Phys. B240 [FS12] (1984) 577.
   C.B. Lang, Phys. Lett. 155B (1985) 399; Nucl. Phys. B265 [FS15] (1986) 630.

3) A. Hasenfratz and P. Hasenfratz Phys. Rev. D34 (1986) 3160.

4) R. Dashen and H. Neuberger, Phys. Rev. Lett. 50 (1983) 1897.
   P. Hasenfratz and J. Nager, Z. Phys. C37 (1988) 477.

5) M. Lüscher and P. Weisz, Nucl.Phys. B290[FS20] (1987) 25; ibid B295[FS21] (1988) 65.
   J. Kuti and Y. Shen Phys. Rev. Lett. 60 (1988) 85.

6) A. Hasenfratz, T. Neuhaus, K. Jansen, Y. Yoneyama and C. B. Lang, Phys. Lett. 199B (1987) 531.
   J. Kuti, L. Lin and Y. Shen preprint UCSD/PTH 87-18
   G. Bhanot and K. M. Bitar, preprint FSU-SCRI-88-50.
   A. Hasenfratz, K. Jansen, J.Jersak,T. Neuhaus, C. B. Lang and Y. Yoneyama, in preparation.

7) W. Langguth and I. Montvay, Z. Phys. C36 (1987) 725
   A. Hasenfratz and T. Neuhaus, Nucl. Phys. B297 (1988) 205.

8. see for example, I.-H. Lee and R. Shrock, Phys. Rev. Lett. 59 (1987)14; Phys. Lett. 196B (1987) 82; Nucl. Phys. B290 [FS20] (1987) 275.
   I.-H. Lee and J. Shigemitsu, Phys.Lett. 178B (1986) 93
   J. Shigemitsu and A. De, OSU preprint DOE-ER-01545-398.

9. J. Shigemitsu, Phys. Lett. 189B (1987) 164; OSU preprint DOE/ER/01545-397.
   I. Montvay, preprint DESY-87-077./hfil/break

10. M. B. Einhorn and G. J. Goldberg, Phys.Rev.Lett. 57 (1986) 2115.

11. J. Polonyi and J. Shigemitsu, OSU preprint DOE-ER-01545-403

12. D. Stephenson and A. Thornton, Edinburgh preprint 88/436.

FOUR DIMENSIONAL, ASYMPTOTICALLY FREE NON-LINEAR SIGMA MODELS

Peter Hasenfratz

Institute of Theoretical Physics
University of Berne
Sidlerstr. 5
3012 Berne, Switzerland

ABSTRACT

The standard non-linear $SU(N)$ sigma-model is not renormalizable in $d=4$, since the dimension=4 operators, which control the ultraviolet behaviour at the Gaussian fixed-point, are absent. By adding a complete set of dimension=4 operators, the model becomes perturbatively renormalizable. Using the background field method the 1-loop beta-functions are calculated and demonstrated that the model is asymptotically free. The dimension=4 operators contain 4 derivatives and the model has negative metric excitations in perturbation theory. However, the couplings become strong in the infrared and the spectrum is a non-perturbative problem. The possible scenarios concerning the fate of the indefinite metric excitation are demonstrated on the exactly solvable $O(N=\text{infinite})$ non-linear sigma-model.

THE ABELIAN HIGGS MODEL IN AN EXTERNAL FIELD

Urs M. Heller

Institute for Theoretical Physics
University of California
Santa Barbara, California 93106

Poul H. Damgaard

The Niels Bohr Institute
Blegdamsvej 17
DK-2100 Copenhagen
DENMARK

ABSTRACT

We study the lattice regularized Abelian Higgs model coupled to an external electromagnetic field. We find a variety of phenomena, expected from the analogy to the phenomenological Ginzburg-Landau theory of superconductivity, such as the Meissner effect and symmetry restoration in sufficiently strong external fields. Finally we present numerical evidence for Nielsen-Olesen vortices under appropriate conditions.

1. INTRODUCTION

It is always interesting to impose a new "extreme" environment on a physical theory, such as high temperature or strong external fields, with properties radically different from the ordinary vacuum. This can lead to a better understanding of several aspects of the theory; it might be realizable in the laboratory (like the plasma phase of QCD), or it may have implications on models of the early universe on a cosmological scale.

Here we study scalar QED, the Abelian Higgs model, in an external electromagnetic field, described by the Euclidean action:

$$S = \int d^4x \left\{ \frac{1}{4}(F_{\mu\nu} - F_{\mu\nu}^{ex})^2 + |D_\mu\phi|^2 + m^2|\phi|^2 + \lambda|\phi|^4 \right\}. \tag{1}$$

From the analogy with the 3-dimensional phenomenological Ginzburg-Landau model of superconductivity we expect a variety of interesting behaviors of this model in the "broken" or Higgs phase:[1] symmetry restoration for sufficiently strong external fields, expulsion of weaker fields—the relativistic analogue of the Meissner effect—and the appearance of Nielsen-Olesen vortices[2] under appropriate conditions. All these phenomena already occur on the (semi-) classical level.[3] Some have also been treated on the 1-loop level in the Coleman-Weinberg limit of perturbation theory.[4] However, several important questions, such as the location of the phase boundary between the Higgs model equivalent of type-I and type-II phases, have not been answered beyond the "classical" level of equations of motion. Here we want to do a full, non-perturbative, treatment of the quantum theory. To this end we regularize the theory by putting it on a space-time lattice, and simulate it by Monte Carlo methods. We shall now review some of the results of these simulations.[5]

2. THE LATTICE MODEL AND SIMULATION METHOD

We latticize the Abelian Higgs model in the by now standard way:

$$S = \beta \sum_p \left[1 - \cos(\vartheta_p - \vartheta_p^{ex}) \right] - \kappa \sum_{x,\mu} \left(\phi_x^* e^{i\vartheta_\mu(x)} \phi_{x+\mu} + \text{h.c.} \right)$$
$$+ \lambda \sum_x (|\phi_x|^2 - 1)^2 + \sum_x |\phi_x|^2, \tag{2}$$

where $\beta = \frac{1}{e^2}$. The above choice for including the external field, though like any lattice action not unique, is conceptually simple and allows us to parallel the usual classical treatment of the model on the lattice as closely as possible. In particular, we want the external field ϑ_p^{ex} such that there are lattice configurations $\{\vartheta_\mu(x)\}$ with $\vartheta_p = \vartheta_p^{ex}$ (mod 2π). On an L^4 lattice with periodic boundary conditions this restricts us (in the constant field sector) to external fields

$$\vartheta_p^{ex} = \frac{2\pi}{L^2} n_{ex}, \qquad n_{ex} \in Z \tag{3}$$

corresponding to $2\pi n_{ex}$ total flux through the planes spanned by the plaquettes p:

$$e\Phi = \sum_{p \in \text{plane}} \bar{\vartheta}_p. \tag{4}$$

Here the reduced plaquette variables $\bar{\vartheta}_p$ are forced to lie in the interval $(-\pi, \pi]$ through the identification[6]

$$\bar{\vartheta}_p = \vartheta_p + 2\pi n_p \tag{5}$$

with $n_p \epsilon \{-2, -1, 0, 1, 2\}$. Note that (4) is the lattice analogue of the flux in the continuum:

$$\Phi_c = \int_S d^2 \sigma_{\mu\nu} F_{\mu\nu}, \tag{6}$$

with analogous expressions for the external flux.

Because of the periodic boundary conditions, we have

$$e\Phi = 2\pi \sum_{p \in \text{plane}} n_p. \tag{7}$$

The reduced plaquettes variable $\bar{\vartheta}_p$ carry the physical flux and give a conserved monopole current

$$M_\mu(x) = \frac{1}{2\pi} \epsilon_{\mu\nu\gamma\delta} \Delta_\nu \bar{\vartheta}_{\gamma\delta}(x), \qquad \Delta_\mu M_\mu = 0. \tag{8}$$

The integers n_p count the number of Dirac strings passing through the given plaquette. The Dirac strings sweep out Dirac sheets, which are either bounded by a closed monopole loop or wrap around the finite, periodic lattice. For each 2π of flux through a lattice plane there is a corresponding Dirac sheet on the orthogonal plane on the dual lattice.

The presence of Dirac sheets wrapping around the lattice causes severe metastability problems in Monte Carlo simulations,[7] since as topological objects they are difficult to create or destroy in local MC updating procedures. This would be particularly devastating for us, since we need to be able to efficiently probe configurations with different total flux. Fortunately, it is easy to construct "classical" lattice configurations of a given flux.[5] We use such configurations to propose, from time to time, global changes that would change the total flux by $\pm 2\pi$, and accepting or rejecting them with a usual Metropolis criterion.

3. MONTE CARLO RESULTS

One classical solution for the lattice Higgs model in an external field is

$$\vartheta_p = \vartheta_p^{ex} \qquad \text{and} \qquad |\phi_x|^2 = 0 \tag{9}$$

corresponding to a symmetric phase. In the Coulomb phase, which contains a massless photon, we would not expect large deviations from this prediction. In particular, we would expect the measured flux to equal the externally applied flux. Indeed, starting from an equilibrium configuration without flux and simulating in a non-zero external field, the system quickly adjusts to the new environment by accepting global changes with the appropriate flux as shown in figure 1. All other quantities, like $E_p = \langle 1 - \cos \vartheta_p \rangle$ for plaquettes p not in the plane of the external flux, $\langle |\phi|^2 \rangle$, etc., remain basically unchanged. Hence we conclude that in the Coulomb phase the external flux penetrates, i.e., the measured flux equals the applied flux.

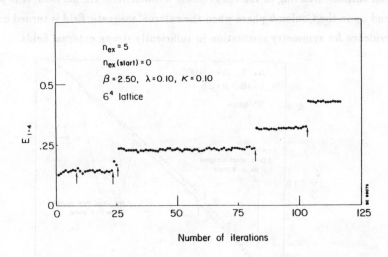

Figure 1. The run history of an equilibrium configuration with no flux from the Coulomb phase made to relax in an external flux $n_{ex} = 5$. The arrows indicate where flux changes have been accepted to adjust to the new environment.

It is interesting to repeat this "experiment" in the strong-coupling confining phase of the model. Here we find that the external magnetic field does not penetrate the lattice. Instead the measured magnetic flux varies essentially randomly in both magnitude and direction as we shift between different equilibrium configurations. On average the magnetic flux tends to vanish. This curious phenomenon gives further support to the hypothesis of "magnetic flux condensation" being the driving force behind confinement in this model.[6] Magnetic monopole-antimonopole pairs populate the vacuum in this sector, and the probability of creating or destroying Dirac sheets in all directions should be very high. This has indeed been clearly demonstrated.[5]

Next we concentrate on the Coulomb-Higgs phase transition. We do so mainly by considering "thermal cycles" in κ at fixed λ and β for different n_{ex} (i.e., different external fields). An example is shown in figure 2. At some points we checked the transition region with longer runs. Monitoring $\langle|\phi|^2\rangle$ or $\langle\phi^* U_\mu \phi\rangle$, not order parameters but rather reliable indicators for the Higgs transition, we find that the κ_c increases with increasing n_{ex}. Hence, when an external field is applied, the Higgs region shrinks. Starting in the Higgs phase *without* the external field, it is possible to end up in the Coulomb phase when the external magnetic field is turned on. This is evidence for symmetry restoration in sufficiently strong external fields.

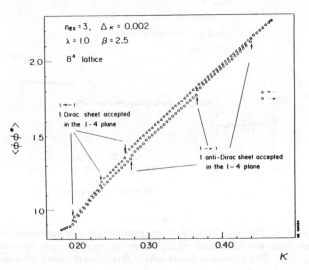

Figure 2. A thermal cycle probing for the Coulomb-Higgs phase transition. Also seen are stepwise jumps that coincide with changes of the measured flux by 2π.

We also observe, for the parameter values that we looked at, a rather long hysteresis in these variables, as well as in E_{1-4}, with stepwise jumps. These stepwise jumps coincide with changes of the measured flux through 1-4 planes by 2π. Going deep enough into the Higgs phase, the flux eventually vanishes, i.e., the external flux is completely expelled: the Meissner effect.

We have found empirically that the point at which the total flux vanishes, for a given n_{ex}, is surprisingly well-defined. It even, at least qualitatively, agrees with the Ginzburg-Landau criterion for the boundary between type I and type II phases:

$$k = \frac{1}{\sqrt{2}} \frac{M_H}{M_W} \Big|_{tree\ level} = \frac{\sqrt{\lambda\beta}}{\kappa}$$

$$k < \frac{1}{\sqrt{2}} \qquad \text{type I} \tag{10}$$

$$k > \frac{1}{\sqrt{2}} \qquad \text{type II.}$$

But obviously, once the flux is expelled, we can only say that we are either in a type I situation with the external field below the critical field B_c or in a type II situation with $B_{ex} < B_{c_1}$.[1]

However, we notice that there is an intermediate region in the Higgs phase, where at least partial flux penetration is observed. This is the first crude indication of a "mixed phase" populated by Nielsen-Olesen vortices.

4. VORTICES

Here we want to see more direct evidence of vortices in the above mentioned intermediate region. In the continuum such vortices carry a quantized magnetic flux, associated with the topological charge of two-dimensional $U(1)$ gauge field configurations on surfaces orthogonal to the vortex world sheet.

On the lattice, we have already shown that the flux is quantized. But for a vortex it should, up to quantum fluctuations, also be localized. Hence, when we want to reveal the existence of vortices, we have to "strip-off" the quantum fluctuations. We do this, on some given equilibrium configuration, by a standard cooling technique, which will bring the configuration to the "closest" lying saddle point. The result reveals nice vortices and even clearly separated vortex pairs, as shown in figures 3 and 4. From the fall-off away from the core[3]

$$B(r) \sim r^{-\frac{1}{2}} e^{-r/\lambda_{p.d.}}; \quad \lambda_{p.d.} = \frac{1}{M_W}$$

$$\text{as } r \to \infty \tag{11}$$

$$|\phi(r)| \sim \phi_0 - \text{const.}\ e^{-r/\xi}; \xi = \frac{\sqrt{2}}{M_H}$$

we see that the $M_H > M_W$, as we would expect in the type II phase.

110

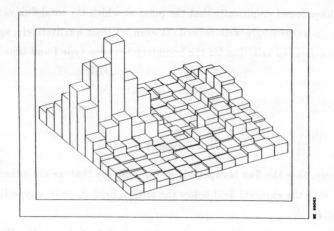

Figure 3. The distribution of the local magnetic flux density on a 12^4 lattice (with periodic boundary conditions) after cooling of a configuration with one unit of flux from the intermediate (type II) Higgs region.

We should of course also mention that we have checked that after cooling:

i) in the Coulomb phase the flux is equally distributed over the whole plane and $|\phi(r)|$ is flat (and "small").

ii) deep in the Higgs phase the flux is zero and $|\phi(r)|$ is flat (and "large").

5. HIGHER CHARGES

The compact version of scalar QED that we are studying on the lattice has one obvious generalization, one where the scalar field transforms under $U(1)$ gauge transformations as a higher "representation" than the fundamental. It simply corresponds to giving the scalar field a charge $Qe, Q = 2, 3, \ldots$ instead of the basic unit e. In our lattice action this is accomplished by the simple replacement

$$\phi_x^* e^{i\vartheta_\mu(x)} \phi_{x+\mu} \mapsto \phi_x^* e^{iQ\vartheta_\mu(x)} \phi_{x+\mu}. \tag{12}$$

In the continuum this theory can be mapped to the usual theory with $Q = 1$ by a change of variables, but with our compact lattice variables this is not the case. Hence here we have an interesting situation where the analogy with the non-

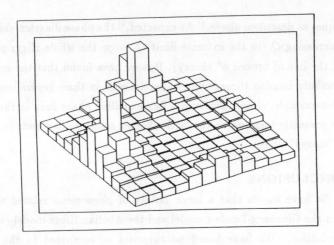

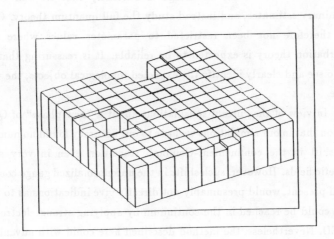

Figure 4. (a) same as in fig. 3 except now with two units of flux.
(b) Corresponding distribution of the Higgs modulus $\rho = \sqrt{\phi^*\phi}$.

relativistic Ginzburg-Landau theory no longer holds directly, and one could imagine that the response to an external electromagnetic field might be different.

We have recently started investigating this question by the same Monte Carlo techniques as described above.[8] As expected,[9] the phase diagram changes radically with increasing Q (in the extreme limit $Q \to \infty$ the whole Higgs phase collapses to just the line of broken ϕ^4 theory). But we have found that the main features of the Ginzburg-Landau theory seem to persist even in these generalized theories. We have, for example, clearly observed a flux-expelling phase deep in the Higgs region. We are presently focusing on the lattice vortex solutions one finds in the equivalent of the "mixed" phase.

6. CONCLUSIONS

We have shown that a large variety of phenomena related to the analogy between the Ginzburg-Landau model and the Abelian Higgs model can be analyzed on the lattice. We have found no surprises as compared to the semi-classical expectations. The main advantage is obviously that we can go beyond such preliminary estimates, and instead study the full quantum theory. Our simulations have therefore not been restricted to parameter values where weak coupling perturbation theory is expected to be reliable. It is reassuring that we have been able to see and clearly identify the expected topological objects, the vortices, on the lattice.

In view of the current speculations about a "new phase" of QED, we should mention that in our simulations we have to our accuracy found no numerical evidence of a shift in the confinement-Coulomb transition even in very strong external magnetic fields. However, such shifts in the unrenormalized gauge coupling constant, even if present, would presumably not directly give indications as to whether such a phase could be reached in the continuum by applying strong electromagnetic fields in QED. Nevertheless, the method described here could with advantage be used to investigate this issue.

REFERENCES

1. "Superconductivity" (R.D. Parks, Ed.), M. Dekker Inc., New York, 1969; D. Saint-James, E.J. Thomas and G. Sarma: "Type II Superconductivity," Pergamon Press, New York, 1965.

2. H.B. Nielsen and P. Olesen, Nucl. Phys. B61 (1973) 45.

3. B.J. Harrington and H.K. Shepard, Nucl. Phys. B105 (1976) 527; D.A. Kirzhnits and A.D. Linde, Ann. Phys. (N.Y.) 101 (1976) 195; A.D. Linde, Rep. Progr. Phys. 42, (1979) 389.

4. A. Salam and J. Strathdee, Nucl. Phys. B90 (1975) 203;
   G.M. Shore, Ann. Phys. (N.Y.) 134 (1981) 259.

5 P.H. Damgaard and U.M. Heller, Phys. Rev. Lett. 60 (1988) 1246;
   and preprint NBI-HE-88-19 and NSF-ITP-88-27i, (1988).

6. T.A. DeGrand and D. Toussaint, Phys. Rev. D22 (1980) 2478.

7. V. Grösch et.al. Phys. Lett. 162B (1985) 171.

8. P.H. Damgaard and U.M. Heller, work in progress.

9. K.C. Bowler et al., Phys. Lett. 104B (1981) 481;
   D.J.E. Callaway and L.J. Carson, Phys. Rev. D25 (1982) 531;
   J. Ranft, J. Kripfganz and G. Ranft, Phys. Rev. D28 (1983) 360.

BREAKING OF CONFINING STRING IN THE SU(2) HIGGS MODEL

J. Jersák

Institute for Theoretical Physics E, RWTH Aachen
D-5100 Aachen, W. Germany
and
HLRZ c/o KFA Jülich, D-5170 Jülich, W. Germany

K. Kanaya

Institute for Theoretical Physics
University of Bern
CH-3012, Bern
Switzerland

ABSTRACT

We analyze the Wilson loops and the gauge invariant 2-point functions by means of an ansatz consisting of two terms, one of them obeying the perimeter law and the other one the area law. In a narrow strip around the Higgs phase transition we find that both terms are important for description of the data obtained on a 16^4 lattice, whereas only one of these terms is sufficient outside of this region. This indicates that in this region a breaking of the confining string is observed. We determine here the values of the string tension, the screening energy of external sources, the renormalized fine structure constant and the order parameter distinguishing between phases with and without confinement.

1. INTRODUCTION

The lattice SU(2) Higgs model with the doublet scalar field Φ,

$$S(\kappa) = -\frac{\beta}{4} \sum_P Tr\left(U_P + U_P^\dagger\right) - \kappa \sum_x \sum_{\mu=1}^4 Re\left(Tr\Phi_x^\dagger U_{x,\mu}\Phi_{x+\mu}\right)$$
$$+ \lambda \sum_x \frac{1}{2}Tr\left(\Phi_x^\dagger\Phi_x - 1\right)^2 + \sum_x \frac{1}{2}Tr\left(\Phi_x^\dagger\Phi_x\right), \tag{1}$$

has been used intensively both for the study of the nonperturbative properties of the Higgs mechanism and for the investigation of the confinement phenomena in gauge theories with dynamical matter fields. These two aspects of the model are complementary and should be understood and taken into account simultaneously when the model is considered within the context of the Glashow-Salam-Weinberg theory of electro-weak interactions. In addition, the experience with the screening of the confining static potential by the scalar field in this model might increase our understanding of the process of hadronization in the QCD.

As far as we know, the model possesses only one phase in the space of the couplings λ, β and κ, and the SU(2) charges are confined everywhere in this phase.[1] The absence of free charges manifests itself in various regions of the coupling space in different ways, however.

For $\kappa = 0$ the Wilson loops of arbitrarily large size decay exponentially with the area of the loops and the static potential rises linearly to infinity. For nonvanishing but small κ this area law behaviour is present only at nonasymptotic distances and changes into an exponential decay with perimeter at distances above a very large screening length at which the confining string breaks. On a lattice of a finite size L^4 only the nonasymptotic area law behaviour of the Wilson loops can be observed, however. We call such a region of the coupling space the *area law (AL) region*. For sufficiently large κ the asymptotic perimeter law behaviour of the Wilson loops can be seen at distances well below $L/2$. The static potential has at all distances, at least approximately, the form of the Yukawa potential. The confined charges are screened by a condensate of the scalar field and the confining string is not present at any distance. Such a region in the coupling space will be denoted the *perimeter law (PL) region*.

Both the AL and the PL regions are understood theoretically to the same extend as we understand the confinement in pure gauge theories or the screening of a charge in a plasma, respectively, and this understanding is nowadays quite good. But for sufficiently large L there is a narrow strip between these regions, called the *string breaking (SB) region*, where at distances smaller than $L/2$ both the linearly rising static potential and its flattening to a constant can be observed simultaneously. In the language of QCD the hadronization, or a breaking of the confinement string, is observable here. Our aim was to localize the three regions and to develop a suitable description of the complex physical phenomena in the SB region.

Working on a lattice of the size $L = 16$ we have investigated[2] the Wilson loops $W(T, R)$ defined on rectangles $T \times R$ and the gauge invariant 2-point functions

$$G(T, R) = \left\langle Tr\Phi_x^\dagger \left(\prod_{\iota \epsilon \Gamma_G} U_\iota \right) \Phi_y \right\rangle, \tag{2}$$

defined on the paths Γ_G connecting the points x and y along 3 sides of rectangles $T \times R$ with $T = |x - y|$. The data have been obtained at about 80 points close to the Higgs phase transition. We have introduced an ansatz for both these functions which is a superposition of two terms, one with the area law behaviour and the other with the perimeter law behaviour. The relative importance of these terms in the fits of our data for $W(T, R)$ and $G(T, R)$ at many points in the coupling space allows us to estimate the position of the AL and PL regions, when the perimeter or the area law term, respectively, is unimportant, and of the SB region, where both terms in the ansatz are needed for an appropriate description of the data.

It turned out that the SB region includes a part of the Higgs phase transition line, which makes the investigation of breaking of the confining string relevant also for the study of the Higgs model in the scaling region. Furthermore, the ansatz makes it possible to extract from the data in the SB region the values of some observables which characterize the area or perimeter law terms in spite of the fact that the nonasymptotic and asymptotic features are present in the data simultaneously. Thus we have been able to estimate in the SB region the string tension σ, the screening energy of the external sources μ and the order parameter ρ_{FM}^∞ introduced by Fredenhagen and Marcu.[3]

2. SUPERPOSITION FORMULAE

2.1. Analytic ansatz for the Wilson loops

Apart from a traditional analysis of the Wilson loops by means of some formula for the static potential[2,4,5] we have looked for an analytic ansatz directly for the Wilson loops. As we want to take into account the perimeter law decay and the area law decay of $W(T, R)$ simultaneously, we have chosen a superposition of two terms with the corresponding decay properties. Combining this with a simple ansatz for the short distance behaviour we arrive at the following *superposition formula for* $W(T, R)$:

$$
\begin{aligned}
W(T, R) = &\left\{ \overline{W}_P \exp\left[-\mu 2\,(T + R)\right] \right. \\
&+ \overline{W}_A \exp\left[-\sigma T R - E_{ext} 2\,(T + R)\right]\Big\} \\
&\times \exp\left[\frac{3}{4}\alpha \left(\frac{T}{R} \exp\left(-m_Y R\right) + \frac{R}{T} \exp\left(-m_Y T\right)\right)\right] \\
\equiv &\,W_P\,(T, R) + W_A\,(T, R)\ .
\end{aligned}
\tag{3}
$$

The first term is the perimeter law term dominant for large distances, and the second one is the area law term which might dominate at intermediate distances. The common Yukawa potential factor represents the short distance properties of $W(T, R)$. As the perimeter and area law terms are symmetric with respect to an interchange of T and R, and as both variables T and R are equally important, we have symmetrized also the short distance part. The list of the 7 parameters defining this formula is:

μ: screening energy of an external charge in the PL region

E_{ext}: selfenergy of an external charge in the AL region

σ: string tension in the AL region

α: renormalized fine structure constant

m_Y: Yukawa mass

$\overline{W}_P$: coefficient of the perimeter law term

$\overline{W}_A$: coefficient of the area law term.

We have indicated the standard physical interpretation of the parameters μ, E_{ext} and σ within the corresponding regions. An extension of this interpretation to the SB region will be discussed later. The ansatz (3) is meant to be valid for $W(T, R)$ for finite T and R, and not intended for determining $V(R)$ in the limit $T \to \infty$. Such an extrapolation from the data obtained on a finite lattice would be too unreliable, as a small T-dependence of $\ln W(T, R)/T$ cannot be excluded.

The symbols $1/\widehat{R}$ and $1/\widehat{T}$ denote the lattice Coulomb potential. We do not use the complete lattice Yukawa potential since it depends on m_Y and would have to be recalculated numerically many times during the fits with m_Y being a free parameter.

2.2. Analytic ansatz for $G(T, R)$

Our ansatz for $G(T, R)$ as a superposition of an area and a perimeter law term is analogous to that for the Wilson loops. The *superposition formula for* $G(T, R)$ is

$$
\begin{aligned}
G(T, R) = \big\{ & \overline{G}_P \exp[-\mu(T + 2R)] \\
& + \overline{G}_A \exp[-\sigma T R - E_{ext}(T + 2R) - \epsilon_c T] \big\} \\
& \times \exp\left[\frac{3}{4} \alpha \frac{R}{T} \exp(-m_Y T) \right] \\
\equiv & \, G_P(T, R) + G_A(T, R) \ .
\end{aligned}
\tag{4}
$$

Most parameters appear already in the formula (3) for $W(T, R)$. The new parameters are

ϵ_c :   bare mass + selfenergy of the constituent particle in the AL region

$\overline{G}_P$:   coefficient of the perimeter law term

$\overline{G}_A$:   coefficient of the area law term.

The formula (4) reproduces again the expected behaviour both at asymptotic and at intermediate distances. The choice of the short range term has some ambiguities which are discussed in Ref. [2].

2.3. Strategy of the fit procedures

We fit $W(T, R)$ and $G(T, R)$ *simultaneously*. Thus the total number of free parameters is 10. Even with some necessary cuts for small R and T[2], the number of data points to be fitted, allowed by the restriction $T, R \leq 8$ on our 16^4 lattice, is about 90. The ratio of the number of data points to the number of free parameters is therefore quite high. On the other hand, we may expect that the data for different R and T are not completely statistically independent, as they are obtained from the same ensemble of configurations. This may effect the errors of the fit parameters.

The χ^2 function is a quadratic form of the 4 fit parameters $\overline{W}_P$, $\overline{W}_A$, $\overline{G}_P$ and $\overline{G}_A$. Therefore it is possible to determine the minimum with respect to these 4 quantities analytically. Doing this one gets an expression for χ^2, which depends only on the remaining 6 parameters, the data and their statistical errors. Thus for the actual MINUIT procedure a tolerable number of 6 free parameters remains. The values of the fit parameters are quite stable with respect to various changes of the fit procedures, described in detail in Ref. [2], and so we believe that they are at least semiquantitatively reliable. However, the error bars produced by the MINUIT program seem to be sometimes unrealistic. The best estimate of the errors can be obtained from the variation of the values of the parameters at close κ-points.

The most important result of the superposition fits is the determination of the κ-values for which both perimeter and area law terms in the superposition formulae are *approximately of the same importance* for the description of the data. This we take as a *signal that the system is in the SB region*. The relative importance of the perimeter and area law terms changes with T and R and it is necessary to devise some "global" comparison of both terms. This is described in Ref. [2].

Of course, the fits produce always some values of the parameters for the perimeter and area terms even in the cases when these terms are unimportant, i.e., in the AL region or the PL region, respectively. *These values of the parameters are not to be believed.* The reason is that the non-dominant terms must be expected to simulate systematic uncertainties in the ansatz for the dominant term, which are unknown. On the other hand, we expect that the physical interpretation of various parameters in the AL and PL regions can be extended into the SB region.

3. RESULTS OF THE ANALYSIS BY MEANS OF THE SUPERPOSITION FORMULAE

Our calculations have been performed at many κ in the vicinity of the Higgs phase transition for $\lambda = 0.5$ and $\beta = 2.1, 2.25, 2.4, 2.6, 3.5$. For $\beta = 3.5$ the string tension is not observable, similarly to the pure SU(2) gauge theory ($\kappa = 0$). But for the other β-values we have been able to determine the boundaries between the AL, SB and PL regions on a 16^4 lattice. The SB region forms a very narrow strip with the width of about 0.007 in κ, the Higgs phase transition at $\kappa = \kappa_{PT}$ being approximately in the middle of this strip. This has been determined using several methods of comparison of the relative importance of the terms $W_P(T, R)$

and $W_A(T,R)$ in eq.(3), and $G_P(T,R)$ and $G_A(T,R)$ in eq.(4) for the description of the data.[2]

In the AL region below the SB region it is not possible to determine the asymptotic properties of $W(T,R)$ and $G(T,R)$ whereas this is easy to do so in the PL region above the SB region. The superposition formulae allow us to extract also from the data in the SB region at least semiquantitative values of the parameters determining both the perimeter and the area law behaviour of these observables.

3.1. κ-dependence of various fit parameters

The results of the superposition fits for the ratios $\overline{W}_A/\overline{W}_P$ and $\overline{G}_A/\overline{G}_P$ are shown in Fig. 1. Three of the parameters with a physical interpretation, namely σ, E_{ext}, and μ are displayed in Fig. 2. In this figure we have shadowed for each of the parameters that κ-region in which the values of the parameter have been obtained from the smaller of the two terms in the superposition formulae. These smaller terms might be strongly influenced by theoretical uncertainties of the ansatz and the values of their parameters are spurious. Therefore *we attribute no physical significance to the values of the parameters in the shadowed regions* and show them only for completeness.

The most remarkable aspect of the Figs. 1 and 2 is that the ratios of the coefficients of the area and perimeter terms change strongly when passing from the AL region via the SB region to the PL region, whereas the values of the three displayed parameters change only moderately. Also the values of the parameter ϵ_c are κ–independent, being approximately $\epsilon_c \simeq 1.5$ for all β-values. Thus the change of the relative importance of both terms in the superposition formulae is mainly due to the change of the coefficients without much change of the physical parameters.

The values of α determined at $\beta = 2.4$ and 2.6 by means of the superposition formulae are quite κ-independent and therefore we just give their values $\alpha = 0.2 \pm 0.02(\beta = 2.4)$ and $\alpha = 0.19 \pm 0.01(\beta = 2.6)$ (for $\beta = 2.1$ and 2.25 a reliable determination of α-values from our data was not possible). A determination of the Yukawa mass from $W(T,R)$ and $G(T,R)$ turned out to be very unprecise.

3.2. Order parameter for confinement

The close relationship between the perimeter laws for $G(T,R)$ and $W(T,R)$ in confining theories with dynamical matter was a motivation for the construction

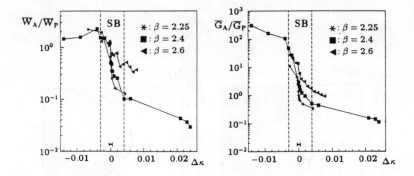

Figure 1. Dependence of the ratios $\overline{W}_A/\overline{W}_P$ and $\overline{G}_A/\overline{G}_P$ in the superposition formulae on $\Delta\kappa = \kappa - \kappa_{PT}$ for three different values of β. Approximate position of the SB region is indicated. Notice the fast decrease of both ratios in the SB region.

of certain "order parameters" distinguishing between confinement and free charge phases of various models. The parameter proposed by Fredenhagen and Marcu[3] is

$$\rho_{FM}^{\infty} = \lim_{T\to\infty} \rho_{FM}(T,R) , \quad R = \text{const. } T , \tag{5}$$

where

$$\rho_{FM}(T,R) = \frac{G(T,R/2)}{W(T,R)^{1/2}} . \tag{6}$$

The parameter ρ_{FM}^{∞} is nonzero in the confinement phase and vanishes in a free charge phase.[3] Therefore ρ_{FM}^{∞} should be nonzero in our model for any $\kappa > 0$.

The asymptotic value ρ_{FM}^{∞} can be calculated from the superposition formulae fits,

$$\overline{\rho}_{FM}^{\infty} = \frac{\overline{G}_P}{\overline{W}_P^{1/2}} , \tag{7}$$

or determined by means of an independent fit to the data for $\rho_{FM}(T,R)$. The results are consistent.[2]

122

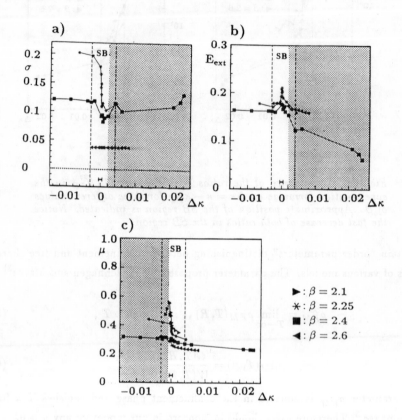

Figure 2. *Values of the parameters* σ, E_{ext}, ϵ_c *and* μ. *No physical meaning can be attributed to the values of these parameters in the shadowed regions. Approximate position of the SB region is indicated.*

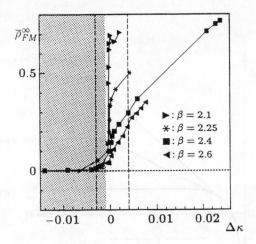

Figure 3. The asymptotic values $\overline{\rho}_{FM}^{\infty}$ of $\rho_{FM}(T,R)$ calculated by means of Eq. (7). The values in the shadowed part are not trustworthy because $W(T,R)$ and $G(T,R)$ are not asymptotic in the AL region.

The values of $\overline{\rho}_{FM}^{\infty}$ are displayed in Fig. 3. We find that the region where this parameter seems to vanish is actually the AL region where it is not possible to determine the asymptotic values of $\rho_{FM}(T,R)$. As this function does not need to depend on T and R monotonously, it is well possible that the order parameter (5) of Fredenhagen and Marcu is, as expected, nonzero also in the AL region. .

4. VANISHING OF THE CONFINING STRING WITH DECREASING DYNAMICAL MASS

One of the most interesting observables is the string tension σ. It decreases slightly in the SB region when κ increases above the boundary between the AL and SB regions but remains finite even at the boundary between the SB and PL regions. A tentative interpretation of this behaviour is illustrated schematically in Fig. 4 if one identifies the string tension σ with the slope of the linear part in the static potential (Fig. 4a). As the extension l of this linear part decreases with increasing κ monotonously, we can consider σ as a function of l instead of κ. In Fig. 4b three possible different scenarios how the linear part of the potential could vanish

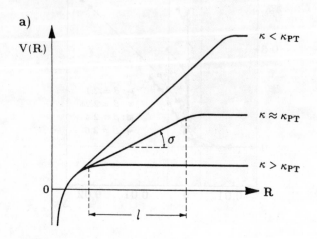

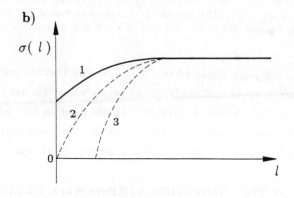

Figure 4. a) *The qualitative shape of the static potential suggested by the the superposition formulae fits in three κ regions.* b) *The slope of the linear part in the static potential as a function of the extension l of this linear part. Three possible behaviours are indicated. The curve 1 corresponds to the scenario suggested by the superposition formulae fits.*

with growing κ are indicated. Our results suggest that in the case of the $SU(2)$ Higgs model this happens according to the curve 1. This means that the area law behaviour of $W(T, R)$ and $G(T, R)$ at intermediate distances disappears because the R- and T-interval of its validity shrinks to zero near the boundary between the SB and PL regions and not because the string tension vanishes at this boundary.

Acknowledgements

We thank the coauthors of this investigation for collaboration and the BMFT and the DFG for financial support. The calculations have been performed at the computer centers of the University of Georgia and the Bochum University. This work was done in collaboration with W. Bock, H.G. Evertz, K. Jansen, H.A. Kastrup, D.P. Landau, T. Neuhaus and J.L. Xu. Some preliminary results have been reported by K.K. during the Seillac Conference, Sept. 1987.

REFERENCES

1. J. Jersák, Lattice Gauge Theory - A Challenge in Large Scale Computing, eds. B. Bunk, K.H. Mütter, and K. Schilling (Plenum Press, 1986) for a review of the properties of the SU(2) Higgs model.

2. W. Bock, H.G. Evertz, K. Jansen, J. Jersák, K. Kanaya, H.A. Kastrup, D.P. Landau, T. Neuhaus, and J.L. Xu, Screening of the confinement potential in the SU(2) lattice gauge theory with matter, Aachen preprint PITHA 88/14.

3. K. Fredenhagen and M. Marcu, Commun. Math. Phys., 92, (1983) 81.
   K. Fredenhagen and M. Marcu, Phys. Rev. Lett., 56, (1986) 223.

4. H.G. Evertz, V. Grösch, J. Jersák, H.A. Kastrup, D.P. Landau, T. Neuhaus and J.L. Xu, Phys. Lett., 175B, (1986) 335.
   H.G. Evertz, V. Grösch, K. Jansen, J. Jersák, H.A. Kastrup and T. Neuhaus, Nucl. Phys., B285 [FS19], (1987) 559.

5. H.A. Kastrup, Nucl. Phys. B (Proc. Suppl.), 4, (1988) 390.

FINITE-SIZE EFFECTS IN THE $SU(2)$ HIGGS MODEL:
A NUMERICAL INVESTIGATION

H. G. Evertz†, E. Katznelson‡*, P. G. Lauwers‡ and M. Marcu†

† II. Institut für Theoretische Physik,
Luruper Chaussee 149, 2000 Hamburg 50,
Federal Republic of Germany

‡ Physikalisches Institut der Universität Bonn
Nußallee 12, 5300 Bonn 1,
Federal Republic of Germany

ABSTRACT

We present new results from a Monte Carlo simulation of the Higgs region of the $SU(2)$ gauge theory with a fundamental scalar matter field. To investigate the model at $\lambda = \infty$, $\beta = 2.4$ and 2.7, and κ close to the phase transition, three different lattice sizes were used: $8^3 \times 24$, $10^3 \times 24$ and $12^3 \times 24$. The finite size effects, which are by no means negligible, are discussed in detail.

1. INTRODUCTION

The nonperturbative properties of the Glashow-Weinberg-Salam model of weak and electromagnetic interactions[1] are extremely important from a purely theoretical as well as from a phenomenological viewpoint. Analytic methods that lead to concrete information about these properties have not been developed yet. The most promising technique has been the Monte-Carlo (MC) simulation of the lattice regularized model. There are however severe conceptual[2] and technical[3] problems in treating the Standard Model fermions on the lattice. On the other hand, today's computers can well handle the problem of gauge fields coupled to scalar matter fields. Before starting a large scale simulation of the full $SU(2) \times U(1)$

---

* speaker at conference

gauge-Higgs system, it is important to gain a good quantitative understanding of the simpler $SU(2)$ and $U(1)$ theories separately.

Here we report on a high-statistics simulation of the $SU(2)$ gauge-Higgs system with the Higgs field in the fundamental representation. Many aspects of this model have been investigated before by other groups, both by simulations[4,5,6,7] and by perturbative methods.[8] From a numerical point of view the model is not an easy one. It has three independent parameters, β, κ, and λ. A complete numerical investigation should include simulations for the full range of these parameters for large enough lattices. While providing an overall picture, the previous simulations often underestimated (or neglected) the dependence of the results on the lattice size. One of the aims of our Monte Carlo study is to deal with the finite size effects in a systematic way.

Some preliminary results for $\beta = 3.0$ and $\lambda = \infty$ were reported on before.[9] We had found considerable finite-size effects in the particle masses for lattices up to $12^3 \times 24$. In order to study these effects in more detail we decided to perform simulations at $\lambda = \infty$ and at smaller values of β, where the W mass is larger. For κ we chose values in the Higgs region ($\kappa > \kappa_{crit}$) close to the phase transition. The strategy is to move in a controlled way towards the region of larger β, where the renormalized weak coupling constant takes values compatible with the experimental one.[7]

In Section 2 we briefly review the model and the physical quantities we studied. In Section 3 we describe the MC simulations and results. Some concluding remarks are presented in Section 4.

2. THE MODEL AND THE MEASURED QUANTITIES

The $SU(2)$ Higgs model, as the $SU(2)$ lattice gauge theory with a Higgs field in the fundamental representation is commonly called, is defined by the action:[4-7,9]

$$S = -\frac{\beta}{2} \sum_P \mathrm{Tr}\, U_P - \kappa \sum_{x,\mu} \rho_x\, \rho_{x+\hat{\mu}}\, \mathrm{Tr}(\sigma_x^{\dagger} U_{x,\mu}\, \sigma_{x+\hat{\mu}}) + \lambda \sum_x (\rho_x^2 - 1)^2 + \sum_x \rho_x^2. \quad (1)$$

Here $U_{x,\mu}$ is the $SU(2)$ element on the link starting from the lattice point $x = (\underline{x}, t)$ in the μ direction, U_P is the usual plaquette variable, $\rho_x = (\phi_x, \phi_x)$ is the length of the two-dimensional complex scalar field ϕ_x, and σ_x is the $SU(2)$ matrix that rotates the vector $(1,0)$ into ϕ_x/ρ_x. As usual, we denote the gauge invariant link variable by

$$V_\mu(x) = \sigma_x^\dagger U_{x,\mu}\,\sigma_{x+\mu} = V_{0,\mu}(x) + i\sum_{r=1}^{3} V_{r,\mu}(x)\,\tau_r, \qquad (2)$$

where τ_r are the usual Pauli matrices. Besides the gauge symetry the model has an $SU(2)$ global "isospin" symmetry (conjugation of $V_\mu(x)$ by a constant matrix). The indices 0 and r in (2) denote the isospin-zero part and the three components of the isospin-one part.

2.1. The W and Higgs masses

To determine the W mass we consider the decay of the correlation function[5–7]

$$K_W(T) = \left\langle \sum_{r=1}^{3}\sum_{j=1}^{3}\left[\sum_{\underline{x}} V_{r,j}(\underline{x},0)\right]\left[\sum_{\underline{y}} V_{r,j}(\underline{y},T)\right]\right\rangle, \qquad (3)$$

where j denotes the three spacelike directions and the summation $\sum_{\underline{x}}$ runs over all three-dimensional space coordinates. For the computation of the Higgs mass an appropriate correlation function is

$$K_H(T) = \left\langle \left[\sum_{\underline{x}}\sum_{j=1}^{3} V_{0,j}(\underline{x},0)\right]\left[\sum_{\underline{y}}\sum_{j=1}^{3} V_{0,j}(\underline{y},T)\right]\right\rangle$$
$$- \left\langle \left[\sum_{\underline{x}}\sum_{j=1}^{3} V_{0,j}(\underline{x},0)\right]\right\rangle^2. \qquad (4)$$

Because of the periodic boundary conditions, the measured correlation functions (3) and (4) are fitted by a periodic exponential decay in T (the fit parameters are A and m):

$$K(T) = A\left\{\exp[-am\,T] + \exp[-am\,(L_t - T)]\right\}. \qquad (5)$$

Here a is the lattice spacing, m is the mass in physical units, and L_t is the lattice size in time direction. Eq. (5) holds for T large, but $T < L_t/2$.

2.2. Static potential, renormalized gauge coupling and source energy

Several interesting physical quantities can be derived by studying the $R \times T$ space-time Wilson loops $W(R, T)$ in the fundamental representation.

The potential $V_W(R)$ between two fundamental representation static sources at distance R is obtained from the exponential decay in time of $W(R, T)$. At fixed R a two-parameter ($C_W(R)$ and $V_W(R)$) fit is performed ($R < T < L_t/2$):

$$W(R, T) = C_W(R) \exp[-V_W(R) T]. \tag{6}$$

After $V_W(R)$ is determined in this way for $1 \leq R \leq L_s/2$ (L_s is the lattice size in space direction), this function of R is compared to the Yukawa potential $Y(R, am_W)$ obtained in the one-W-exchange approximation. For this we do a two-parameter (α_{ren} and E_q) fit:

$$V_W(R) = -\frac{3\alpha_{ren}}{4} Y(R, am_W) + 2E_q. \tag{7}$$

Here either the continuum Yukawa form, $Y_{cont}(R, am_W) = \exp(-am_W R)/R$, or a lattice version $Y_{latt}(R, am_W)$,[6,7] which takes into acount the correct range for the momenta on the finite lattice, can be used. Since this potential is derived as a first perturbative approximation (one W, no loops), it cannot be expected to fit the data well for all values of β, κ, and λ, and for the full range of R. If the fit is good, then α_{ren} is equal to $g_{ren}^2/4\pi$ with g_{ren} the renormalized gauge coupling, and E_q is the energy of a fundamental static source. Depending on the version of the Yukawa potential used, we will write $E_{q,cont}$ and $\alpha_{ren,cont}$, or $E_{q,latt}$ and $\alpha_{ren,latt}$.

In principle one could fit the potential with m_W as a third free parameter, instead of using the value obtained from eq. (5). In practice, however, this is notoriously difficult[5,6,7,9] since for the range of available R-values the fits are rather insensitive to changes in m_W.

The static source energy E_q can also be determined in two other ways. The first method uses the expectation values of gauge invariant timelike lines:

$$G_1(T) = \left\langle \frac{1}{2} \text{Tr} \left[\sigma_{(\underline{x},0)}^{\dagger} \left(\prod_{t=0}^{T-1} U_{(\underline{x},t),0} \right) \sigma_{(\underline{x},T)} \right] \right\rangle. \tag{8}$$

For large values of T but with $T < L_t/2$, a two-parameter (C_1 and $E_{q,line}$) fit is performed:

$$G_1(T) = C_1 \exp(-E_{q,line}T). \tag{9}$$

The second alternative consists of directly computing the source energy from the measured expectation value of the timelike Polyakov loop $P_t \equiv G_1(L_t)$ by means of the relation

$$E_{q,Pol} = -(1/L_t) \ln(P_t). \tag{10}$$

These different methods for computing the same physical quantity provide an excellent way of checking the consistency of measurements and computational methods.

2.3. The Vacuum Overlap Order Parameter (VOOP)

In the continuum theory it is customary to consider the Higgs expectation value $v = <\phi^c>$ (ϕ^c is the continuum Higgs field), which is related to the Fermi coupling constant. This quantity v cannot be defined unless the gauge is fixed. A better quantity to consider is the VOOP. It has a clear gauge-independent physical interpretation.[10] In addition to the Wilson loops $W(R,T)$, another set of gauge-independent quantities $G_2(R,T)$ must be measured:

$$G_2(R,T) = \left\langle \frac{1}{2}\,\mathrm{Tr}\left[\sigma^\dagger_{(\underline{x},0)} \left(\prod_{t=0}^{T-1} U_{(\underline{x},t),0} \prod_{r=0}^{R-1} U_{(\underline{x}+r\hat{1},T),1} \right. \right. \right.$$
$$\left. \left. \left. \times \prod_{t=T-1}^{0} U^\dagger_{(\underline{x}+R\hat{1},t),0} \right) \sigma_{(\underline{x}+R\hat{1},0)} \right] \right\rangle \tag{11}$$

The VOOP is then defined as the limit $R,T \to \infty$ of the function

$$\rho(R,T) = \frac{G_2(R,T)}{W(R,2T)^{1/2}} \tag{12}$$

with $T \geq cR$ (c is a positive constant). It can be shown that for all gauges in the tree-level approximation the VOOP $\equiv \rho(\infty,\infty)$ is related to v by the expression

$$(av)^2 = \kappa\,\rho(\infty,\infty). \tag{13}$$

In the next section we will refer to the VOOP determined in this way as VOOP_{dir}.

It is not always an easy task to determine this limit numerically in the way described before, especially in the case of small lattices. A good way to avoid these difficulties is to take the limit $T \to \infty$ first, while keeping R fixed at a finite value. We can do this for the numerator and the denominator separately by performing two-parameter fits according to eq. (6) and to

$$G_2(R,T) = C_2(R) \exp[-V_{G_2}(R)T]. \tag{14}$$

V_W and V_{G_2} represent the same potential, fitted from two different quantities. If the fits (6) and (14) are both good and after checking that $V_W = V_{G_2}$, we have:

$$\rho(R,\infty) = \frac{C_2(R)}{C_W(R)^{1/2}}. \tag{15}$$

This allows us to determine the VOOP if the lattice is large enough in the space direction. The value obtained in this way will be referred to as VOOP$_{rat}$.

It has been suggested theoretically that in a phase without free charges the VOOP is equal to the constant C_1 obtained from the fit (9).[10] This is a numerically easier way to determine the VOOP. As in the case of the source energy, having more than one method to determine a given quantity provides a useful check of the internal consistency of our calculation.

3. THE MONTE CARLO SIMULATIONS AND THE RESULTS

As opposed to the case of pure matter theories, for the $SU(2)$ Higgs model a theory of the finite size effects has not been worked out. The most reliable way to check whether one has these effects under control still is to perform simulations on lattices of increasing size until the changes in the measured quantities become smaller than the statistical errors. For this reason we carried out simulations for three different lattice sizes ($8^3 \times 24$, $10^3 \times 24$, $12^3 \times 24$). We will refer to these sizes as small, medium and large respectively.

We will present data for $\lambda = \infty$, $\beta = 2.4$ and 2.7. For each β we took two values of κ that are in the Higgs phase, relatively close to the phase transition. The MC runs started with approximately 15000 thermalization sweeps, followed by 90000 to 120000 sweeeps during which measurements were performed every third sweep. We used a fully vectorized version[11] of Creutz's heat-bath algorithm.[12] In order to check whether any flip-flops to the confinement phase had occured, the full history of eight quantities was stored. In the subsequent analysis we found no such flip-flops.

We were very careful with the error analysis. By the central limit theorem,[13] the averages of the measured physical quantities over a long time series are approximately Gaussian random variables. The Gaussian distribution is characterized by a covariance matrix. One way to estimate it is to partition the time series into bins and compute the covariance matrix for the bin averages. As the bin size is increased, this bin average covariance matrix divided by the number of bins converges to the covariance matrix for the whole-run averages. In computing functions of the measured quantities or in fitting them, we did the error analysis using the whole covariance matrix. We will discuss below how the more widespread procedure of only considering the variances (i. e., the diagonal of the covariance matrix) may lead to numerical mistakes.

Our analysis of the MC results is summarized in Tables 1 and 2. Almost all symbols used in these tables were defined in Section 2. The last two entries are the values of the W masses as computed from the tree-level relation (16) (see Section 3.5). The numbers followed by a * are not as trustworthy as the others. They either result from fits that are not good (reduced $\chi^2 > 1$), or from a numerical limiting procedure where it was not clear that the asymptotic value had really been reached.

3.1. The W and Higgs masses

Fitting the masses with eq. (5) is a delicate task. One has to vary the shortest distance T_{min} and the longest distance T_{max} and find the situations for which the reduced χ^2 is smaller than one. The value of the fitted parameters and their statistical error depends very much on the number of data points considered, especially on T_{min}. The results quoted here are those with the smallest errors among the subset of fits that have $\chi^2 < 1$ and that give stable mass values with respect to small changes in T_{min} and T_{max}.

TABLE 1: Collected results for $\beta = 2.7$

Lattice size	$\kappa = .365$			$\kappa = .370$		
	$8^3 \times 24$	$10^3 \times 24$	$12^3 \times 24$	$8^3 \times 24$	$10^3 \times 24$	$12^3 \times 24$
am_W	.41(2)	.32(2)	.33(2)	.40(2)	.31(1)	.31(1)
am_H	.59(2)	.53(3)	.53(3)	.56(8)	.73(2)	.74(2)
$VOOP_{dir}$	.101(2)	.101(4)	.097(2)*	.130(2)	.129(2)	.126(2)
$VOOP_{rat}$	.103(6)	.105(4)	.105(2)	.133(4)	.133(3)	.131(5)
C_1	.101(6)*	.104(4)	.103(3)	.138(5)	.137(2)	.135(1)
$E_{q,line}$	.267(6)*	.251(5)	.243(3)	.257(4)	.242(2)	.235(2)
$E_{q,Pol}$	.277(4)	.258(2)	.243(1)	.256(3)	.241(1)	.234(1)
$E_{q,cont}$	.219(1)	.228(2)*	.228(1)	.218(1)	.225(2)	.224(1)
$E_{q,latt}$	.206(2)	.218(3)	.220(1)	.210(1)	.216(1)	.217(1)
$\alpha_{ren,cont}$	.297(6)	.39(2)*	.36(2)	.296(4)	.37(2)	.37(2)
$\alpha_{ren,latt}$	.279(4)	.36(2)	.34(1)	.276(4)	.34(2)	.32(1)
$am_{W,cont}$	.27(1)	.31(1)*	.29(1)	.30(1)	.34(1)	.33(1)
$am_{W,latt}$	.26(1)	.29(1)	.29(1)	.29(1)	.32(1)	.31(1)

TABLE 2: Collected results for $\beta = 2.4$

Lattice size	$\kappa = .390$			$\kappa = .400$		
	$8^3 \times 24$	$10^3 \times 24$	$12^3 \times 24$	$8^3 \times 24$	$10^3 \times 24$	$12^3 \times 24$
am_W	.47(2)	.41(2)	.38(2)	.47(1)	.44(1)	.40(2)*
am_H	.80(2)	.55(6)*	.67(4)	.91(3)	.96(3)	.94(3)
$VOOP_{dir}$	.165(3)	.167(6)	.168(2)	.217(3)	.216(2)	.216(5)
$VOOP_{rat}$	.172(14)	.168(4)	.171(4)	.216(7)	.215(9)	.215(12)
C_1	.183(3)	.170(3)	.178(2)*	.228(3)	.231(1)	.226(1)*
$E_{q,line}$	.312(3)	.298(2)	.301(2)*	.288(2)	.286(1)	.282(1)*
$E_{q,Pol}$	.306(9)	.301(4)	.297(3)	.290(6)	.281(3)	.279(2)
$E_{q,cont}$	.283(8)*	.293(5)*	.293(4)*	.287(2)	.281(1)	.280(3)
$E_{q,latt}$	.270(5)*	.283(2)*	.284(3)*	.268(1)	.276(1)	.275(2)
$\alpha_{ren,cont}$	.56(12)*	.61(7)*	.58(4)*	.64(4)	.58(3)	.51(8)
$\alpha_{ren,latt}$	.51(11)*	.53(6)*	.50(4)*	.60(3)	.51(3)	.42(7)
$am_{W,cont}$	.48(7)*	.50(3)*	.49(2)*	.59(3)	.56(3)	.52(6)
$am_{W,latt}$	.46(7)*	.47(3)*	.46(2)*	.57(3)	.53(3)	.48(5)

For a correct analysis of the goodness of fit, it is absolutely necessary to use the full covariance matrix for the measurements at different values of T. We compared the fits done in this way with those disregarding the covariances between different quantities ,i. e. , using only the variances. For the same T_{min} and T_{max}, the values of the fitted parameters differed only marginally (rarely up to one standard deviation); the errors estimated with the full covariance matrix were usually smaller by 20 to 50%. On the other hand, the value of χ^2 was often larger for the fits that used the full covariance matrix, in some cases by as much as 100%. Thus if we only use the variances, we may be misled to declare as good a fit starting from a too small T_{min}. In this case it is not only the errors that are not estimated correctly, but also the values of the fitted physical quantities.

For our values of the coupling constants, fitting the Higgs mass is more difficult than fitting the W mass, as the correlation function K_H drops off much faster and becomes zero (within error bars) for distances around 7 lattice units.

The analysis for the data at $\beta = 2.7$ went very smoothly. From Table 1 it is clear that the difference in masses between the medium and large lattice size is zero within error bars. It is interesting that $\kappa = 0.365$ and 0.370 give the same W mass. At $\beta = 3.0$[9] there was a similar situation close to the phase transition. There we could not exclude the possibility that this was due to finite-size effects. For $\beta = 2.7$ we are now more confident that the finite size effects are under control. New data for $\beta = 3.0$ on larger lattices, that will be published elsewhere, point in the same direction.

The data for $\beta = 2.4$ are not as good as for $\beta = 2.7$. As expected for a smaller β, the measurements are much "noisier". This makes fits more difficult, especially for the W mass. Again, its values for $\kappa = 0.39$ and 0.40 agree within error bars. On the other hand it is also clear that the results for the medium and large lattice are still quite different. New data on a $14^3 \times 28$ lattice seem to confirm the values quoted for the large lattice.

The Higgs mass varies strongly with κ, as expected from other studies.[5,6,7,9] In our case, m_H is in the region of $2m_W$. It may be that some of the finite size effects are due to the crossover (as the lattice size is changed) from a situation with a stable Higgs to one where the Higgs particle is a resonance.

3.2. Static potential, renormalized gauge coupling and source energy

As described in Section 2.2, the fit of the potentials to the measured data is done in two steps. First the potentials $V(R)$ for different spatial distances R are determined together with their covariance matrix. This procedure is similar to the one used for the masses and it always worked smoothly. However, employing the full covariance matrix (of $W(R,T)$) makes an even greater difference here.

The second step, where some potential function is fitted to $V(R)$, is more delicate. One of the reasons is that the correct potential function that should be used to fit the measured data is not known theoretically. This leads to the paradoxical situation that it is often easier to fit potentials to lower statistics data than to high statistics data.

At $\beta = 2.7$ the fits with both versions of the Yukawa potential were satisfactory in all cases except one. For distances R in the range $2 \leq R \leq L_s/2$ the fits were always good. Furthermore, the results did not change when the values of the potential for $R = 6$ and/or $R = 5$ were excluded. All results quoted in Table 1 are for $2 \leq R \leq 4$. Only in one case ($\kappa = 0.365$, medium lattice, continuum potential) the fit for $3 \leq R \leq 5$ was quite bad and gave results that were different from the ones quoted in Table 1. For this reason we marked this case by a *.

For our data at $\beta = 2.4$ and $\kappa = 0.390$ the fitting of potentials was not successful for two reasons: the data are noisier and the renormalized gauge coupling is quite large. Usually the fit including only distances from 3 to 5 was good, but these results are not very reliable because the fitted parameters are quite insensitive to the data for large distances. If we include the data for $R = 2$ the fits become bad. The explanation is that for these values of the parameters and for these distances the one-particle-exchange approximation breaks down and different potential functions should be used. Previous simulations[6] at similar values of β also found that close to the phase transition the Yukawa fits do not work well. On the other hand, for $\kappa = 0.400$ the fitting was as good as for the two κ values at $\beta = 2.7$. This is probably due to the fact that by increasing κ, am_H increases and α_{ren} decreases.[5,6] In Table 2 we have again quoted the results for $2 \leq R \leq 4$, even for $\kappa = .390$.

The energy E_q of a static source can be computed in four different ways. On a finite lattice with large L_t, $E_{q,Pol}$ and $E_{q,line}$ should be equal. The finite size corrections to $E_{q,cont}$ and $E_{q,latt}$ however should neither be equal to each other, nor to those of $E_{q,Pol}$ and $E_{q,line}$. This different finite size dependence can indeed be seen from our data. For the small lattice the four values for the source energy are usually somewhat apart. At $\beta = 2.4$, $\kappa = .400$, they converge towards compatible values as the lattice size is increased. At $\beta = 2.7$ however, they are still not equal within error bars on the large lattice.

3.3. VOOP

For the small lattice it is very difficult, if not impossible, to obtain a reliable value for VOOP_{dir} because it is not feasible to check whether the asymptotic value has already been reached. However from the data for the large lattices one clearly sees that this limit has already been reached within error bars for $R = 3$ or 4. For this reason the values and errors quoted for the small lattice in the tables are the ones for $R = 3$.

As described in Section 2.3 there is a more reliable way to determine the VOOP, by taking at fixed R the limit $T \to \infty$ for the numerator and the denominator independently (VOOP_{rat}). For the small lattices the subsequent limit $R \to \infty$ is again problematic but for the larger lattices this procedure gives a stable asymptotic value. A third method to determine the VOOP is to use C_1 (see (9)). In the data C_1 is always larger or equal (within error bars) than the values for the VOOP calculated in the two previous ways. Again the different methods should not have identical finite size effects. For $\beta = 2.7$ they converge towards compatible values. For $\beta = 2.4$ the procedure to determine C_1 was not so successful and the values for C_1 are still larger than VOOP_{dir} or VOOP_{rat}.

It should be noticed that the values that we obtained for VOOP_{rat} hardly change with growing lattice size. This is one additional reason to trust this method more than the other ones.

3.4. Tree-level relations

From our discussion of the VOOP ($\equiv \rho(\infty, \infty)$) it is clear that the gauge invariant two-point function $\rho(R, \infty)$ can be determined numerically in a consistent way in the Higgs region. It is instructive to check the validity of the tree-level relation that gives m_W as a function of $\rho(\infty, \infty)$ and α_{ren}:

$$(a m_W)^2 = 2\pi \, \alpha_{ren} \, (a \, v)^2 = 2\pi \, \alpha_{ren} \, \kappa \, \rho(\infty, \infty). \tag{16}$$

In this relation we can use $\alpha_{ren,latt}$ or $\alpha_{ren,cont}$, and the values for m_W are labeled correspondingly. The values obtained in this way were included in the tables.

The results for $\beta = 2.7$ agree quite well numerically with the values for m_W obtained directly, at least for the larger lattices where we do no longer expect strong finite-size effects. A similar result was found for $\beta = 3.0$.[9] This indicates that the renormalized perturbation theory can still be used here. Notice that the bare and the renormalized weak couplings differ considerably at $\beta = 2.7$. Perturbation theory predicts that at fixed β the r. h. s. of (16) grows with $(\kappa - \kappa_{crit})$. This too seems to be reflected in the behavior of the data. As pointed out before, our values for the directly measured m_W show (within error bars) no κ dependence, both at $\beta = 2.7$ and 3.0.

For $\beta = 2.4$, the tree-level relation for the W mass does not hold , even for the large lattice. This should be expected for small values of β.

4. CONCLUSIONS

We performed a simulation in the Higgs region of the SU(2) lattice gauge theory with a fundamental scalar matter field, close to the transition to the confinement region. We investigated the finite size effects for several quantities: masses m_H and m_W, renormalized gauge coupling α_{ren}, source energy E_q and gauge invariant Higgs expectation value defined by the VOOP.

As we had hoped at the outset, the finite size effects at $\beta = 2.7$ are smaller than they had been at $\beta = 3.0$,[9] and our results here converge to the values on the $12^3 \times 24$ lattice (with the notable exception of E_q). Somewhat surprisingly, there

138

are larger finite size effects at the still smaller $\beta = 2.4$.

In the data presented here, and also at $\beta = 3.0$,[9] we do not see (within error bars) an increase with κ of the directly measured value of am_W close to the Higgs phase transition.

It is remarkable that the tree level relation (16) is fullfilled quite well at $\beta = 2.7$ (and 3.0) although $\alpha_{ren}/\alpha_{bare}$ is large (2 to 3). The values of α_{ren} itself and of λ_{ren} defined by $\lambda_{ren} = m_H^2/4v^2$ are not too large for the renormalized perturbation theory to be applicable. On the other hand, the values of v as computed from the VOOP are very different from the values of v in the pure matter theory at $\lambda = \infty$ and similar κ.[14] Thus we cannot expect that all corrections due to the gauge coupling can be computed using the tree-level approximation. In order to really compare the simulation with perturbation theory, a one-loop calculation with the results expressed in terms of the renormalized couplings is needed.

ACKNOWLEDGEMENTS

We are indebted to Klaus Fredenhagen and István Montvay for many helpful discussions. The simulations were performed on the CYBER 205 of the Bochum University, the CRAY X-MP/4-8 of the "Höchstleistungsrechenzentrum" in Jülich and the IBM 3090 of the GMD in Bonn.

REFERENCES

1. S. Weinberg, Phys. Rev. Lett. 19 (1967) 1264;
   A. Salam in *Elementary Particle Theory: Relativistic Groups and Analyticity*, edited by N. Svartholm, Almqvist and Wiksell (1968), p. 367.

2. J. Smit, in *Field Theory on the Lattice* (Seillac 1987), edited by A. Billoire, R. Lacaze, A. Morel, O. Napoly and J. Zinn-Justin, Nucl. Phys. B (Proc. Suppl.) 4 (1988), p. 451.

3. M. Fukugita, in the Seillac proceedings, p. 105.

4. C. B. Lang, C. Rebbi and M. Virasoro, Phys. Lett. 104B (1981) 294;
   H. Kühnelt, C. B. Lang and G. Vones, Nucl. Phys. B230 (1984) 16;
   J. Jersák, C. B. Lang, T. Neuhaus, and G. Vones, Phys. Rev. D32 (1985) 2761;

W. Langguth and I. Montvay, Phys. Lett. 165B (1985) 135;
V. P. Gerdt, A. S. Ilchev, V. K. Mitrjushkin, I. K. Sobolev and A. M. Zadorozhny, Nucl. Phys. B265 (1986) 145;
D.J.E. Callaway, R. Petronzio, Nucl. Phys. B267 (1986) 253.

5. H. G. Evertz, J. Jersák, C. B. Lang and T. Neuhaus, Phys. Lett. 171B (1986) 271;
   H. G. Evertz, V. Grösch, J. Jersák, H. A. Kastrup, D. P. Landau, T. Neuhaus and J.-L. Xu, Phys. Lett. 175B (1986) 145;
   H. G. Evertz, PhD Thesis, Aachen (1987).

6. I. Montvay, Nucl. Phys. B269 (1986) 170;
   W. Langguth, I. Montvay and P. Weisz, Nucl. Phys. B277 (1986) 11.

7. W. Langguth and I. Montvay, Z. Phys. C36 (1987) 725;
   A. Hasenfratz and T. Neuhaus, Nucl. Phys. B292 (1988) 205.

8. A. Hasenfratz and P. Hasenfratz, Phys. Rev. D34 (1986) 3160;
   I. Montvay, Nucl. Phys. B293 (1987) 479.

9. E. Katznelson, P. Lauwers and M. Marcu, in the Seillac proceedings, p. 427.

10. K. Fredenhagen and M. Marcu, Commun. Math. Phys. 92 (1983) 81,
    Phys. Rev. Lett. 56 (1986) 223, and in preparation;
    M. Marcu, in Lattice Gauge Theory – a Challenge to Large Scale Computing (Wuppertal 1985), edited by B. Bunk, K. H. Mütter and K. Schilling, Plenum (1986) p. 267.

11. K. Fredenhagen and M. Marcu, Phys. Lett. 193B (1987) 486.

12. M. Creutz, Phys. Rev. D21, (1980) 2308.

13. T. W. Anderson, The Statistical Analysis of Time Series, Wiley (1971).

14. A. Hasenfratz, K. Jansen, C. B. Lang, T. Neuhaus and H. Yoneyama, Phys. Lett. 199B (1987) 531.

NON-PERTURBATIVE LATTICE STUDY OF THE HIGGS SECTOR IN THE STANDARD MODEL

Julius Kuti* with Lee Lin and Yue Shen

Department of Physics
University of California at San Diego
La Jolla, CA 92093, USA

ABSTRACT

The theoretical breakdown of the standard electroweak model is investigated when the mass of the Higgs particle is larger than the weak interaction scale. A similar problem is identified for a heavy top quark. Qualitative upper bounds on the Higgs mass and the top quark mass are reproduced from the flows of the renormalization group equations of the Higgs self-coupling and the Yukawa coupling of the top quark at large energy scales. The triviality argument for an upper bound on the Higgs mass is made quantitative on the lattice in a large scale computer simulation of the spontaneously broken Higgs sector. It is shown that the SU(2) Higgs sector with spontaneously broken O(4) symmetry and perturbative gauge coupling is always driven to a trivial Gaussian fixed point as criticality is approached at small, large, or infinite bare self-coupling of the Higgs field. A quantitative upper bound, including our new Monte Carlo results, is reported for the mass of the Higgs particle. We find an upper bound $m_H \approx 640 \ GeV$ on the Higgs mass at a dimensionless correlation length $\xi = 2$ on the lattice. The bound will decrease logarithmically with increasing ξ.

1. QUALITATIVE RENORMALIZATION GROUP ANALYSIS

In the standard electroweak model [1-3] with $SU(2)_L \times U(1)$ symmetry the Higgs sector is described by a complex scalar doublet Φ with quartic self-interaction,

$$\Phi = \begin{pmatrix} \Phi^+ \\ \Phi^0 \end{pmatrix} . \tag{1}$$

---

* speaker at the conference

The Euclidean action of the $SU(2)_L$ symmetric Higgs sector is given by

$$S = \int d^4x \left[\frac{1}{2}\left(\partial_\mu + ig\frac{1}{2}\vec{\tau}\cdot\vec{W}_\mu\right)\Phi^\dagger \cdot \left(\partial_\mu - ig\frac{1}{2}\vec{\tau}\cdot\vec{W}_\mu\right)\Phi + V(\Phi^\dagger\Phi) \right]$$
$$+ S_{gauge} + S_{fermion} + counterterms . \qquad (2)$$

The O(4) symmetric Higgs potential has the form

$$V(\Phi^\dagger\Phi) = -\frac{1}{2}m^2\Phi^\dagger\Phi + \lambda(\Phi^\dagger\Phi)^2 , \qquad (3)$$

where m^2 is a positive mass parameter in the broken symmetry phase and λ designates the quartic Higgs self-coupling. The icon S_{gauge} in Eq. (2) designates the non-Abelian gauge field part of the action with an SU(2) triplet vector field $\vec{W}_\mu$ and gauge coupling g (the notation $\vec{\tau}$ stands for the three isospin Pauli matrices). The SU(2) coupling constant g is expected to generate small perturbative corrections to the dynamics of the Higgs sector.

The fermion masses originate from $S_{fermion}$ through Yukawa couplings. Only heavy fermions whose masses are comparable to the weak interaction scale will require strong Yukawa couplings. Light fermion Yukawa couplings are negligible in the dynamics of the Higgs sector.

Although in Eq. (2) continuum notation is used, we will introduce lattice regularization in our Monte Carlo calculations where the momentum cut-off Λ is equal to π/a with lattice spacing a. Counterterms in Eq. (2) render the mass and coupling constant parameters of the Lagrangian finite in the continuum limit.

1.1. Perturbative triviality of the Higgs coupling

Consider the Euclidean Lagrangian for the Higgs sector before the gauge and Yukawa couplings are turned on,

$$\mathcal{L} = \frac{1}{2}(\partial_\mu\Phi)^\dagger(\partial_\mu\Phi) - \frac{1}{2}m^2\Phi^\dagger\Phi + \lambda(\Phi^\dagger\Phi)^2 + counterterms . \qquad (4)$$

Shifting the origin of the field Φ in the spontaneously broken phase by the vacuum expectation value v, we find that Eq. (4) describes three massless Goldstone bosons (w^+, w^-, z^0) and one neutral Higgs particle h with mass $m_H = \sqrt{2}m$. The Fermi coupling constant G_F and the Higgs boson mass m_H are related to v and λ by

$$\frac{1}{v^2} = \sqrt{2}G_F , \quad \lambda = \frac{G_F m_H^2}{4\sqrt{2}} . \qquad (5)$$

The vacuum expectation value $v \approx 250 \; GeV$ is determined by the Fermi coupling constant and λ has a quadratic dependence on the Higgs mass.

In leading order of the SU(2) coupling constant the gauge vector bosons acquire a mass m_W in the broken symmetry phase,

$$m_W = \frac{1}{2} g \cdot v \quad . \tag{6}$$

When the gauge coupling is turned on, the Goldstone bosons will also acquire gauge-dependent masses through mixing with the longitudinal vector bosons.

The one-loop renormalization group equation (RGE) for the Higgs self-coupling is given by

$$\frac{d\lambda}{dt} = \frac{6}{\pi^2} \lambda^2(t) \, , \tag{7}$$

where t is the logarithm of the variable energy scale. If $\lambda(\mu)$ is the renormalized coupling defined at some physical scale μ, and if $\lambda(\Lambda)$ is the coupling constant defined at the cut-off scale Λ (bare coupling), then the relation

$$\frac{1}{\lambda(\mu)} = \frac{1}{\lambda(\Lambda)} + \frac{6}{\pi^2} \, ln\frac{\Lambda}{\mu} \tag{8}$$

follows from Eq. (7) when integrated between $t = 0$ and $t = ln(\Lambda/\mu)$ with the initial condition $\lambda(t = 0) = \lambda(\mu)$. It follows from Eq. (8) that $\lambda(\mu)$ vanishes in the limit μ fixed and $\Lambda/\mu \to \infty$, suggesting triviality for the renormalized coupling at a finite physical scale when the cut-off is removed from the theory. For fixed and finite cut-off Λ the largest value of $\lambda(\mu)$ for a given scale μ is obtained from Eq. (8) in the limit of infinite bare coupling $\lambda(\Lambda)$.

If the coupling constant $\lambda(\mu)$, at a typical weak interaction scale $\mu = m_W$, is approximately identified with the renormalized coupling λ in Eq. (5), an upper bound is implied by Eq. (8) as a function of the cut-off Λ,

$$\lambda(\mu) = \frac{m_H^2}{8v^2} \leq \frac{\pi^2}{6 \cdot ln\Lambda/\mu} \, . \tag{9}$$

To avoid triviality we have to work at finite cut-off, but the ratio Λ/m_H has to be larger than one as a protection against serious cut-off artifacts. On the lattice, with dimensionless Higgs correlation length $\xi = 1/(a \cdot m_H)$, we can write $\pi \cdot \xi = \Lambda/m_H$ and the smallest acceptable value of ξ will determine an upper bound on the mass of the Higgs particle.

There are three obvious flaws which render the above argument for the upper bound on the Higgs mass qualitative, at best. First, the connection between bare and renormalized coupling in Eq. (8) was derived in perturbation theory which is not applicable for large bare coupling. Second, the precise identification of $\lambda(\mu)$ with $m_H^2/(8v^2)$ at some physical scale μ requires higher order corrections. Third, the smallest acceptable value of ξ, where cut-off artifacts are still small, can only be determined from a detailed scaling study of the theory. Only an exact analytic calculation valid for arbitrary bare coupling, or a large scale Monte Carlo simulation which is free from restrictions on the bare coupling will solve these difficulties.

1.2. Heavy top quark mass

With a similar qualitative argument, based on the one-loop RGE, one can study the influence of a heavy top quark mass on the upper bound for the mass of the Higgs particle. If the top quark is heavy, the associated Yukawa coupling G, which is related to the quark mass by the relation $m_t = \frac{G}{\sqrt{2}} \cdot v$, has to be included in the coupled 1-loop RGE,

$$
\frac{d\lambda}{dt} = \frac{6}{\pi^2} \left[\lambda^2 + \frac{1}{8} G^2 \lambda - \frac{1}{64} G^4 \right] ,
$$
$$
\frac{dG}{dt} = \frac{9}{32\pi^2} G^3 .
\tag{10}
$$

The presence of the negative G^4 term in Eq. (10) displays the destabilizing effect of the negative fermion loop contribution to the effective Higgs potential.

One can now integrate the coupled RGE in Eq. (10) from $t = 0$, with initial conditions $\lambda(t = 0) = \lambda(\mu)$ and $G(t = 0) = G(\mu)$, to $t = ln(\Lambda/\mu)$ which leads to a relation between the pair $\lambda(\mu)$, $G(\mu)$ and the bare couplings. With the choice $\mu = m_W$ one can calculate the maximum allowed value of $\lambda(\mu)$ for fixed $G(\mu)$ and finite cut-off Λ. For the choice $\Lambda/m_W \approx 10$, which is somewhat small according to our Monte Carlo results, the upper bound on the Higgs mass is depicted in Figure 1 as a function of the top quark mass. The dashed line in Figure 1 is the upper bound on the top quark mass as determined from the one-loop vacuum instability of the effective Higgs potential.

There are three important observations one can make from the qualitative RGE results. First, the qualitative Higgs bound derived from RGE is surprisingly close to our exact Monte Carlo results in spite of the above mentioned flaws in the

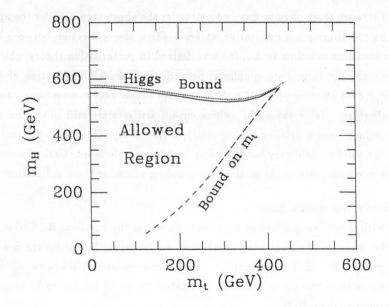

Figure 1. The allowed domain of the Higgs mass and the top quark mass of the SU(2) Higgs sector is determined from one-loop RGE. The solid line depicts the Higgs bound at zero gauge coupling, the dotted line includes g on the one-loop level. The dashed line is the vacuum instability boundary for the effective Higgs potential.

argument. Second, the heavy top quark has little effect on the upper bound for the mass of the Higgs particle, if m_t does not exceed several hundred GeV (in Eq. (10) the QCD corrections to quark loops are neglected which presumably has little effect on the Higgs bound). Third, the SU(2) gauge coupling g in the RGE calculation leads to a minor modification of the Higgs bound in Figure 1. Therefore, one can hope that the O(4) model of Eq. (4) is a very good approximation to the study of the Higgs bound for the full SU(2) sector of the electroweak Lagrangian, if the fermion masses are not heavier than several hundred GeV.

The predictive power of RGE for Higgs and heavy fermion mass bounds was recognized early in the electroweak model, with strong focus on grand unified theories.[4,5] Our qualitative derivation follows the spirit of an influential paper[6] with extensions to the SU(2) gauge coupling and the Yukawa coupling of the heavy fermion.[7,8] We will turn now to the description of our computer simulation results of the O(4) model[9,10] without any perturbative, or other approximations. We will also report on new data obtained in an extended collaboration.[7]

2. NON-PERTURBATIVE TRIVIALITY OF THE O(4) MODEL

The O(4) limit of the SU(2) Higgs model with lattice regularization is given by the Euclidean lattice action[9,10]

$$S = \frac{1}{2} \sum_i \left[\sum_{\hat{\mu}} (\vec{\Phi}_{i+\hat{\mu}} - \vec{\Phi}_i)^2 - m^2 \vec{\Phi}_i^2 \right] + \lambda \sum_i \vec{\Phi}_i^4 , \qquad (11)$$

where, in the absence of counterterms, m^2 and λ are bare lattice parameters, and the field variables $\vec{\Phi}_i$ are O(4) vectors on lattice sites labelled by subscript i. The unit vector $\hat{\mu}$ points along the four positive lattice directions (the lattice spacing is set to unity in our calculations, unless indicated otherwise).

Consider the phase diagram of the O(4) symmetric Euclidean lattice action of Eq. (11) in the bare parameter space of λ and m^2. A critical line, where the inverse Higgs mass diverges in lattice spacing units, will separate the spontaneously broken Higgs phase from the symmetric phase. The non-perturbative demonstration of the trivial Gaussian fixed point requires to show that by tuning m^2 towards its critical value, the logarithmic decrease $\lambda_R \to 0$ is observed for any fixed λ as we approach the critical line. We will use a recently developed non-perturbative technique[11] to study the constraint effective potential for the verification of the conjectured triviality scenario and the related upper bound.

2.1. Effective Potential

The constraint effective potential $U_\Omega(\vec{\phi})$ is defined by[11-13]

$$\exp[-\Omega U_\Omega(\vec{\phi})] = \int D[\vec{\Phi}] \, \delta(\vec{\phi} - \frac{1}{\Omega} \sum_i \vec{\Phi}_i) \, e^{-S[\vec{\Phi}]} , \qquad (12)$$

where Ω designates the finite lattice volume (it is equal to the number of lattice sites in our units) and the integration is over the field variables $\vec{\Phi}_i$. The O(4) symmetric effective potential $U_\Omega(\vec{\phi})$, which is non-convex in the broken symmetry phase and depends on $\phi = \sqrt{\vec{\phi}^2}$ only, has a direct physical interpretation: the probability density $P(\vec{\phi})$ to find the system in a state of average "magnetization" $\vec{\phi}$ is given by $P(\vec{\phi}) = const \cdot e^{-\Omega U_\Omega(\vec{\phi})}$, in close analogy with statistical physics. With this unique feature we can develop a direct and visual physical picture of spontaneous symmetry breaking.

The critical line in the bare parameter space is observed at the crossover points of the effective potential from convex to non-convex shape as m^2 is tuned for fixed λ. Although there is no conventional order parameter in a finite volume for spontaneous symmetry breaking of the continuous $O(4)$ symmetry, the effective potential provides a clean signal for the transition. Zero curvature along three orthogonal directions on the $O(3)$ symmetric manifold of minimum points for the non-convex effective potential $U_\Omega(\phi)$ is associated with the existence of three massless Goldstone particles in the broken symmetry phase which requires careful treatment in the Monte Carlo analysis.

From $U_\Omega(\vec{\phi})$ we can extract the Euclidean Green's functions at zero momentum. For a one-component scalar field one finds the relation

$$\frac{1}{\Omega^n} \sum_{i_1,\ldots,i_n} \langle \Phi_{i_1} \ldots \Phi_{i_n} \rangle = \frac{\int d\phi \, \phi^n \, e^{-\Omega U_\Omega(\phi)}}{\int d\phi \, e^{-\Omega U_\Omega(\phi)}} \tag{13}$$

for the n-point Euclidean Green's function at zero momentum.

For large Ω, the integral in Eq. (13) can be evaluated in saddle point approximation where the moments of ϕ are determined from the derivatives of the effective potential at its minimum ϕ_{min}. We find the well-known relations

$$\lim_{\Omega \to \infty} \frac{\partial^2 U_\Omega(\phi)}{\partial \phi^2}\bigg|_{\phi=\phi_{min}} = \frac{m_R^2}{Z_\Phi} , \tag{14}$$

$$\lim_{\Omega \to \infty} \frac{\partial^4 U_\Omega(\phi)}{\partial \phi^4}\bigg|_{\phi=\Phi_{min}} = 24 \frac{\lambda_R}{Z_\Phi^2} , \tag{15}$$

where Z_Φ is the wavefunction renormalization of the Φ field, m_R^2 is the renormalized mass square in lattice spacing units and λ_R is the renormalized quartic coupling constant. The definitions of the renormalization constant Z_Φ and the renormalized mass m_R are given by

$$\partial \tilde{G}(p^2)^{-1}/\partial p^2 \big|_{p^2=0} = Z_\Phi^{-1} ,$$
$$\tilde{G}^{-1}(0) = m_R^2/Z_\Phi , \tag{16}$$

where $\tilde{G}(p^2)$ is the two-point connected Green's function in momentum space.

The generalization of the relations in Eqs. (13-16) to the four-component scalar field is straightforward. Some care has to be exercised with the infinite volume limit, since the massless Goldstone modes in the broken symmetry phase generate infrared singularities at zero momentum. The finite lattice volume acts as infrared regulator and physical results can be extracted after careful analysis.[7,14]

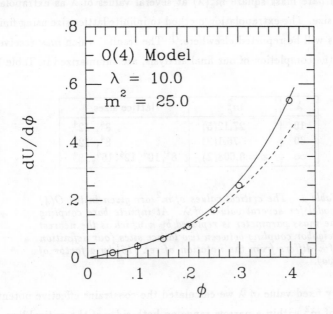

Figure 2. The derivative of the effective potential in the O(4) model is shown in the symmetric phase at lattice size 8^4.

The derivative of the effective potential is measured in a constraint Fourier space Hybrid Monte Carlo simulation as explained elsewhere.[9,11] Figure 2 shows a typical result for the constraint effective potential in the symmetric phase. The dashed line is the tree level value of the potential as determined from a fit of the second and fourth derivatives of $U_\Omega(\phi)$ at $\phi = 0$ from data for small ϕ values. The renormalized Higgs mass, from our fits to the effective potential and the propagator, is $m_H = 0.609(5)$ in lattice units, and the renormalized coupling is $\lambda_R = 1.15(2)$. The wavefunction renormalization constant $Z_\Phi = 0.996(2)$ is obtained from the slope of the propagator in momentum space. The solid curve of Figure 2 is the one-loop form of the effective potential using the fitted values of m_H and λ_R as input parameters in the renormalized loop expansion. Further details of this calculation are given elsewhere.[7,14]

2.2. The trivial Gaussian fixed point

From a large number of Monte Carlo runs we first determined accurate values for the critical bare mass square $m_c^2(\lambda)$ at several values of λ as extrapolated to infinite lattice size. The extrapolation method to infinite lattice size using finite size scaling analysis will be reported elsewhere.[7] The results, which may receive small corrections at the completion of our final analysis, are summarized in Table 1.

λ	m_c^2	lattice sizes
10	27.12(5)	8^4, 12^4
30	76.1(2)	8^4, 12^4
∞	$\kappa = 0.608(2)$	$8^4, 10^4, 12^4, 16^4, 18^4$

Table 1   The critical values of m^2 are given in the $O(4)$ model for several values of λ. At infinite bare coupling the mass parameter is replaced by κ which is the nearest neighbor coupling between two lattice sites (our definition of κ differs from the notation of ref. [16] by a factor of two).

For every fixed value of λ we calculated the constraint effective potential at several values of m^2 within a narrow range on both sides of the critical line. From the constraint effective potential we determined the renormalized coupling constant λ_R in the symmetric phase as m^2 is tuned towards its critical value for fixed λ. At $\lambda = \infty$ we measured connected four-point functions in momentum space. Figure 3 shows some of our most important results for λ_R at two different bare couplings.

The renormalized coupling λ_R is shown as a function of $\tau = 1 - m^2/m_c^2(\lambda)$ at $\lambda = 10.0$ and $\lambda = \infty$. We should note that $\lambda = 10$ corresponds to strong bare coupling in our normalization of the quartic interaction term of the Lagrangian. The solid line is a fit to the form $\lambda_R = const \cdot \left| ln|\tau| \right|^{-\eta}$, where the exponent η is derived from RGE analysis of a trivial Gaussian fixed point,[15] with $\eta = 1$. We find $\eta = 1.03(8)$ in our fit,[10] with a constant amplitude of 3.0(1).

We believe that the existence of the trivial Gaussian fixed point at $\lambda_R = 0$ is strongly supported by the results of Figure 3 where, for strong bare coupling,

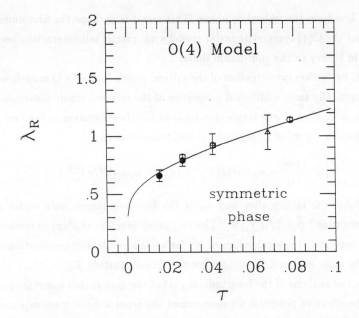

Figure 3. The renormalized coupling is shown in the symmetric phase of the O(4) model. The list of the symbols is given in Table 2 for the identification of data points according to coupling and lattice size. Two new points at infinite bare coupling were added to our earlier results (ref. [10]) by measuring connected four point functions in momentum space (ref. [7]). Although the new data have lower accuracy, they are in very good agreement with our previous results based on our high accuracy effective potential technique.

symbol	λ	lattice size
open circle	10	8^4
full circle	10	12^4
open triangle	∞	8^4
full triangle	∞	12^4
open diamond	∞	∞

Table 2   Symbols are shown for the O(4) data for two values of λ and different lattice sizes. The diamond symbol for ∞ lattice size will be used for results based on finite size scaling extrapolation.

renormalization effects make λ_R small and logarithmically decreasing as we approach the critical line from the unbroken phase. This result is perhaps the first numerical evidence that the O(4) symmetric scalar model with quartic self-interaction becomes a trivial field theory in the continuum limit.

Next, by further investigation of the critical region in the O(4) model, we will explore numerically some additional properties of the infrared stable Gaussian fixed point at $\lambda_R = 0$. One of our important tools in this investigation is the two-point function which is a sum of a longitudinal and transverse part,

$$\left[G^{-1}\right]_{ij}^{\alpha\beta} = n_\alpha n_\beta (G_L)_{ij}^{-1} + (\delta_{\alpha\beta} - n_\alpha n_\beta)(G_T)_{ij}^{-1} \, , \qquad (17)$$

where i, j label the lattice sites and n_α is the four-component unit vector along the "magnetization" $\vec{\phi} = 1/\Omega \sum_i \vec{\Phi}_i$. The two-point function $\tilde{G}_L^{-1}(p)$ in momentum space was measured at several values of m^2 for an independent determination of the renormalized mass m_H and the field renormalization constant Z_Φ.

From the analysis of the longitudinal part of the momentum space propagator and from the effective potential we determined the renormalized mass m_H and the field renormalization constant Z_Φ in the symmetric and broken phases, the vacuum expectation value of the renormalized field operator in the broken symmetry phase, and the renormalized coupling constant λ_R in the symmetric phase.

The renormalized mass m_H is plotted against $\tau = 1 - m^2/m_c^2$ in Figure 4. Assuming a Gaussian fixed point at $\lambda_R = 0$, the logarithmic corrections to mean field critical behavior at the higher critical dimension d=4 were calculated earlier.[9,10,15] The solid line in the unbroken phase is a fit to the scaling form with calculable logarithmic correction, $m_H = const \sqrt{|\tau|} \left| \ln |\tau| \right|^{-\frac{1}{4}}$. The value of the constant is 2.79(4) from our fit. In the broken phase the solid line is our fit for $\lambda = \infty$ with an amplitude of 3.97(3). Three data points extrapolated to ∞ lattice size were used in the fit. The other two points are subject to further finite size analysis. The dashed line is a fit to $\lambda = 10$ data with an amplitude of 4.66(3). Three data points were used in the fit, the last point at $m_H = 1.0$ is subject to further analysis (some changes may occur in our final error analysis). The wavefunction renormalization constant Z_Φ was found to be close to 1 (within a few percent) from the slope of our momentum space propagators.

The vacuum expectation value of the renormalized field operator, identical to the average absolute magnetization $\langle \phi_R \rangle$ in the language of statistical physics, is

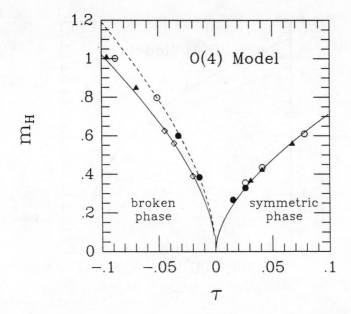

Figure 4. The renormalized mass in the critical region is shown in the O(4) model. The couplings and lattice sizes can be identified from Table 2. The errors are comparable, or smaller than the size of the symbols.

shown in Figure 5 as a function of the τ variable. The solid line is the logarithmically corrected scaling law, $\langle \phi_R \rangle = const\sqrt{|\tau|}\left|\ln|\tau|\right|^{\frac{1}{4}}$ in the spontaneously broken phase with a fitted amplitude of 0.823(1) at $\lambda = \infty$. The three data points extrapolated to ∞ lattice size were used in the fit, the last two points are subject to further finite size analysis. The dashed line at $\lambda = 10$ has the same τ dependence with an amplitude of 0.94(2). We should note again that some changes may occur in our final error analysis.

The logarithmic scaling violation terms are consistent everywhere with our data points, as shown in Figures 3-5. It is particularly significant that we observe the correct logarithmic behavior of the renormalized coupling constant in the symmetric phase (Figure 3). This logarithmic term is *not* a small correction to a rapidly decreasing mean field function, as skeptics might note about the rapid decrease of m_H, or $\langle \phi_R \rangle$, in the critical region. In Figure 3 we determined the exponent η of the logarithmic scaling violation term from our data and excellent agreement was

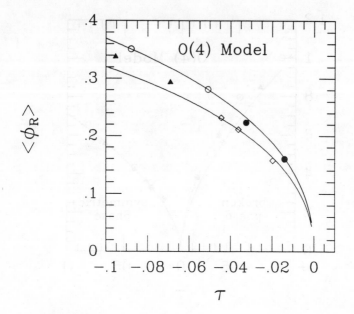

Figure 5. The vacuum expectation value of the renormalized field operator is shown in the O(4) model. The couplings and lattice sizes can be identified from Table 2. The errors are comparable, or smaller than the size of the symbols.

found with the prediction of RGE on a trivial Gaussian fixed point. In Figures 4 and 5 we fitted the data to their mean field theory form corrected for logarithmic scaling violation terms with fixed exponents, as given by the trivial Gaussian fixed point to one-loop order.

In summary, we believe that strong evidence has been presented from our Monte Carlo simulation results to confirm the expectation that the critical behavior of the O(4) scalar field theory is governed by a Gaussian fixed point at $\lambda_R = 0$.

3. TRIVIALITY BOUND ON THE HIGGS MASS

It is often advocated that the Supercollider will be built to find the Higgs particle. According to a "No-Lose Corollary" either a light Higgs particle with a mass much less than 1 TeV will be produced and studied directly, or the Higgs sector will become strongly interacting and new physics will be observed on the TeV energy scale. The corollary originates from a perturbative high energy theorem[17] and a related unitarity bound[18] which were discussed elsewhere, in relation to the O(4) approximation.[9] The corollary, however, may require some modification with the newly explored critical properties of the O(4) model. The modification is intimately related to the triviality bound on the mass of the Higgs particle.

In the language of modern quantum field theory, our computer simulation results strongly indicate that the complex Higgs doublet of the SU(2) sector in the standard electroweak model is defined on a trivial Gaussian fixed point in the continuum limit. The renormalized coupling λ_R has to vanish then in the infinite cut-off limit. At finite cut-off it becomes difficult, if not impossible, to realize the second alternative of the "No-Lose Corollary" by tuning the bare parameters to a strongly interacting Higgs sector without loosing the low-energy effective theory which has only negligible cut-off artifacts. As we will explain now, the theory probably will break down completely before it ever became strongly interacting with a very heavy Higgs particle.

As we mentioned earlier, triviality in the infinite cut-off limit implies an upper bound on the mass of the Higgs particle. Consider the phase diagram of the O(4) symmetric Lagrangian of Eq. (4) in the bare parameter space with lattice regularization. The critical line, where the inverse Higgs mass diverges in lattice spacing units, will separate the spontaneously broken Higgs phase from the symmetric phase. The trivial Gaussian fixed point implies that the renormalized coupling λ_R vanishes as we approach the critical line for any fixed bare λ. For any reasonable definition of the renormalized coupling, the relation

$$\frac{M_H^2}{v^2} = 8\lambda_R + O(\lambda_R^2) \tag{17}$$

holds in perturbation theory. In leading order of the gauge coupling the relation $m_W = \frac{1}{2}gv$ holds between the renormalized vector boson mass and the renormalized Higgs field expectation value v, so that m_H/m_W will have to vanish on the critical line (continuum limit) as $\lambda_R \to 0$.

To support a finite physical Higgs mass in the theory, we have to keep the cut-off finite in physical units and stay away from the critical line at $\tau = 0$. The ratio m_H/m_W will grow with increasing τ, but we cannot push the ratio very high without leaving the critical region. At that point cut-off artifacts begin to dominate physical results and a reasonably well defined upper bound should be identified for the ratio m_H/m_W at the boundary of the scaling region.

154

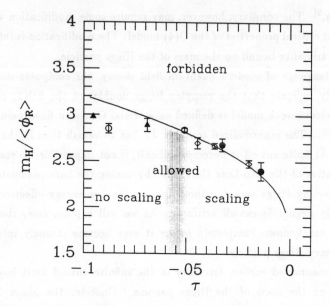

Figure 6. The triviality upper bound on the Higgs mass is illustrated with our new data on the $m_H/\langle\phi_R\rangle$ ratio.

Our data on the Higgs bound and the related physical picture are depicted in Figure 6. At infinite bare coupling λ we fitted the ratio $m_H/\langle\phi_R\rangle$, which is directly related to m_H/m_W by the relations $\langle\phi_R\rangle = v$, $m_W = \frac{1}{2}gv$, to a logarithmic function of τ as required by the trivial Gaussian fixed point for small values of τ. The solid line is a fit to the form $m_H/\langle\phi_R\rangle = const \left|\ln|\tau|\right|^{-\frac{1}{2}}$, with the fitted value of 4.86(4) for the constant. Since for smaller values of λ the ratio data are lower, the solid line separates the allowed and forbidden regions of $m_H/\langle\phi_R\rangle$ in the theory. We observe that $\lambda = 10$ is already quite close to $\lambda = \infty$. The data on $m_H/\langle\phi_R\rangle$ are breaking away from the logarithmic solid line in the grey area (represented by dots) where cut-off artifacts may become significant.

Inside the scaling region we have a reasonable effective theory with nonvanishing renormalized coupling λ_R. At a dimensionless Higgs correlation length $\xi = 2$ we find $m_H/\langle\phi_R\rangle = 2.6$ which corresponds to $m_H \approx 640$ GeV in physical units. For smaller correlation lengths scaling violations become clearly visible. A more detailed study of the scaling violation terms would be desirable to decide whether $\xi = 2$ is the appropriate dividing line between the two regions.

4. CONCLUSION

We established a Higgs mass upper bound of $m_H \approx 640 \; GeV$ at a dimensionless Higgs correlation length of 2 in the $O(4)$ approximation to the $SU(2)$ sector of the electroweak model. There are several ambiguities in determining an "accurate" value for the Higgs bound in the $O(4)$ model.

First, we do not know in a very quantitative way the onset of scaling violation terms with strong cut-off artifacts for Higgs correlation lengths $\xi < 2$. There has been a recent suggestion that perhaps a sensible theory can be extended to larger Higgs masses with strongly interacting Higgs dynamics.[19] We find this scenario unlikely.[20]

Second, there is some concern about the universality of the results. What is the sensitivity of the Higgs bound to the regularization scheme in the non-perturbative calculation?[21] In a stimulating early paper on the Higgs mass bound, momentum space regularization scheme was used in approximate non-perturbative study of RGE in the $O(4)$ model[22] with physical results which are approximately in the same range as our Monte Carlo data.

We believe that the non-perturbative study of an improved lattice action of the $O(4)$ model along the Symanzik program[23] would provide further useful information on the question of regularization scheme dependence, because we would be closer to Euclidean invariance. The real cut-off in the theory is provided by the threshold of the new physics, and our regularization scheme should resemble it as closely as possible.

There are many interesting projects left for the future. One could study the Higgs bound in the standard model with two Higgs doublets. Perhaps more challenging will be an accurate determination of the top quark mass bound based on the instability of the Higgs effective potential in the presence of heavy fermion loop contributions and the triviality problem of the Yukawa coupling.

In closing, we should remark that remarkable progress has been made recently[24] to calculate the Higgs bound in the $O(4)$ model from a large order high temperature expansion continued to the broken phase through the critical line.[25] Detailed comparison of our work with the new analytic results will be very interesting.

A Monte Carlo calculation of the upper bound on the Higgs mass was reported at this conference with results similar to ours.[26,27] Some results on the upper bound were also reported earlier in the full SU(2) Higgs model.[28,29] There have been also some other attempts to study the effective potential in scalar field theories using Monte Carlo techniques.[30-32] Our results are different from those attempts in a number of significant ways.

Acknowledgements

We would like to thank the organizers for their hospitality and the stimulating workshop atmosphere at the conference. This research was supported by DOE grant DE-AT03-81-ER40029 and by a DOE grant of supercomputer time at SCRI, and an NSF grant of supercomputer time at the San Diego Supercomputer Center.

REFERENCES

1. S. Glashow, Nucl. Phys., 22, (1961) 579.

2. S. Weinberg, Phys. Rev. Lett., 19, (1967) 1264.

3. A. Salam, Elementary Particle Theory, eds. W. Svartholm (Almquist and Wiksell, Stockholm, 1968).

4. L. Maiani, G. Parisi, R. Petronzio, Nucl. Phys., B136, (1978) 115.

5. N. Cabibbo, L. Maiani, G. Parisi, R. Petronzio, Nucl. Phys., B158, (1979) 295.

6. R. Dashen and H. Neuberger, Phys. Rev. Lett., 50, (1983) 1897.

7. J. Kuti, L. Lin, S. Meyer, Y. Shen, U.C. San Diego preprint UCSD/PTH 88-08, July 1987.

8. M. Lindner, Z. Phys., C31, (1986) 298.

9. J. Kuti, L. Lin and Y. Shen, Nucl. Phys. (Proc. Suppl.), B4, (1988) 397.

10. J. Kuti, L. Lin and Y. Shen, Upper Bound on the Higgs Mass in the Standard Model, U.C. San Diego preprint UCSD/PTH 87-18, November 1987 (to appear in Phys. Rev. Letters).

11. J. Kuti and Y. Shen, Phys. Rev. Lett., 60, (1988) 85.

12. R. Fukuda and E. Kyriakopoulos, Nucl. Phys., B85, (1975) 354.

13. L. O'Raifeartaigh, A. Wipf and H. Yoneyama, Nucl. Phys., B271, (1986) 653.

14. Y. Shen with J. Kuti, L. Lin, and S. Meyer, Renormalized Perturbation Theory and New Simulation Results in the O(4) Model, this Proceedings, U.C. San Diego preprint UCSD/PTH 88-06, June 1988.

15. E. Brézin, J. C. Le Guillou, J. Zinn-Justin, Field Theoretical Approach to Critical Phenomena, in: Phase Transitions and Critical Phenomena, eds. C. Domb and M. S. Green, (Academic Press, London, 1973) volume 6.

16. M. Lüscher and P. Weisz, Nucl. Phys., 290 [FS20], (1987) 25.

17. M. S. Chanowitz, Universal W,Z Scattering Theorems and No-Lose Corollary for the SSC, in: Proceedings of the XXIII International Conference on High Energy Physics, eds. S. C. Loken, (World Scientific, 1987) 445.

18. B. W. Lee, C. Quigg, and H. B. Thacker, Phys. Rev., D16, (1977) 1519.

19. M. B. Einhorn and D. N. Williams, On the Lattice Approach to the Maximum Higgs Boson Mass, this Proceedings, University of Michigan, Ann Arbor preprint, May 1988.

20. L. Lin with J. Kuti and Y. Shen, How Far is N=4 from N=infinity?, this Proceedings, U.C. San Diego preprint UCSD/PTH 88-07, June 1988.

21. G. Bhanot and K. Bitar, Regularization Dependence of the Lattice Higgs Mass Bound, this Proceedings, FSU-SCRI-88-50 preprint, June 1988.

22. P. Hasenfratz and J. Nager, Zeit. Phys., C37, (1988) 477.

23. K. Symanzik, Nucl. Phys., 226, (1983) 187.

24. P. Weisz, Scaling Laws and Triviality Bounds in the Lattice N Component ϕ^4 Theory, this Proceedings.

25. M. Lüscher and P. Weisz, Nucl. Phys., B295 [FS21], (1987) 65.

26. A. Hasenfratz, K. Jansen, C. B. Lang, T. Neuhaus, and H. Yoneyama, Phys. Lett., 199B, (1987) 531.

27. W. Langguth and I. Montvay, Z. Phys., C36, (1987) 725.

28. A. Hasenfratz and T. Neuhaus, Nucl. Phys., B297, (1988) 205.

29. C. Whitmer, Princeton University Ph. D. Thesis, unpublished, 1983.

30. M. M. Tsypin, Lebedev Institute preprint no. 280, June 1985.

31. K. Huang, E. Manousakis and J. Polonyi, Phys. Rev., D35, (1987) 3187.

MONTE CARLO STUDY OF THE O(4) Φ^4 MODEL

C. B. Lang

Inst. f. theoretische Physik
Universität Graz
A-8010 Graz
AUSTRIA

ABSTRACT

Some recent results of a direct Monte Carlo study of the four-component Φ^4 model in the broken phase on lattices up to 16^4 are presented.

1. MOTIVATION

Most of the work to be presented here evolved from a collaboration with Anna Hasenfratz from SCRI, Tallahassee, Karl Jansen, Jiri Jersák, and Hiroshi Yoneyama, all from the RWTH Aachen, FRG, and Thomas Neuhaus from the Universität Bielefeld, FRG. First results have been presented last year[1] and a complete account is in preparation.[2] I want to thank my friends for the possibility to present our results at this meeting.

When one started to do Monte Carlo (MC) studies of the coupled system of gauge fields and scalar Higgs bosons the main idea was to find out the mechanism by which the vector states become massive in a non-perturbative way. The Higgs field in the Lagrangian is gauge variant and thus it is obvious that the full theory cannot follow the textbook scenario of spontaneous symmetry breaking: A local symmetry cannot be broken spontaneously. As noted already in the beginning[3] the gauge invariant operators building the observable particle spectrum will be bilinear in the original boson field.

MC studies of the phase diagram of, e.g., the $SU(2)$ Higgs model have shown that the confinement region and the so-called Higgs region are connected in the space of couplings. This duality explains that the observed spectrum of bound states is the same in both regions. For the non-abelian gauge group $SU(2)$ the scalar and vector states are massive in both regions with minimal values at the PT.[4-7] For the abelian gauge group $U(1)$ one finds a massless vector state (the photon) in the Coulomb phase, in addition to massive scalar and vector states (bosonium states).[8-9]

Most of the high statistics studies of lattice Higgs models relied on two simplifications:

- One does not take into account the fermions; this is justified by the assuption that fermions do not play a crucial *rôle* for the mechanism of symmetry breaking and mass generation. Some recent studies, however, indicate other interesting features when fermions are included.[10]

- One considers only a subgroup like, e.g., $U(1)$ or $SU(2)$; the justification here again lies in the expectation that the main effects may be understood from the study of a subsystem.

These mentioned MC investigations attempt to study both main concepts: (a) what becomes out of spontaneous symmetry breaking, (b) the Higgs mechanism of the gauge bosons acquiring mass. In tree level perturbation theory we may characterize these "ingredients" by the relations

$$m_H^2 = 4\kappa\lambda < \Phi^2 >, \qquad (1a)$$

$$m_W^2 = \frac{2\kappa}{\beta} < \Phi^2 > . \qquad (1b)$$

The notation follows the usual definition of the lattice action.[4,8] One of the results of these studies was that the *gauge coupling* renormalizes weakly and remains of the order of size of its bare value whereas there are indications that the renormalized *quartic coupling* may differ substantially from its bare value. There is no *a priori* reason to expect that relation (1b) should hold at values of β as small as 2.25 or, depending on the gauge group, $g^2 \simeq O(1)$. However, the non-perturbative results for m_W^2 and $< \Phi^2 >$ appear to follow this relation in the Higgs region of the phase diagram (cf. Fig. 1) remarkably close. A similar comparison based on (1a) for the Higgs mass is unsuccessful and a possible reason may be a significant renormalization of the quartic coupling λ.

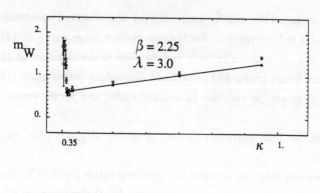

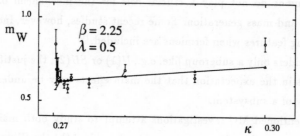

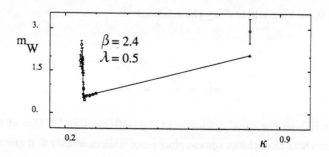

Figure 1. The values of m_W (circles) obtained in Ref. [5] compared with the values obtained from eq.(1b) and the measured expectation value $< \Phi^2 >$ (triangles),(cf. also Ref. [9]). The bare values of $\beta = \frac{4}{g^2}, \kappa$ are used in the relation. A similar figure for the $U(1)$ lattice Higgs model may be found in Ref. [8].

There is ample evidence, that the pure scalar part of the model, the Φ^4 theory, is free in the limit of vanishing lattice spacing.[11] Formulating this statement in the language of spin models: the only fixed point of the theory is of Gaussian nature. At any point in the broken phase, away from the PT, the correlation length is finite, corresponding to a non-vanishing lattice spacing or a finite momentum cut-off. Simulating the lattice theory at such a point leads to an effective theory of bosons with renormalized mass m^R and quartic coupling λ^R, which will be bound from above by a finite value depending on the cut-off.[12,13] This bound is independent on the bare coupling and decreases towards zero in the approach towards the PT point, the continuum limit. This non-perturbative effect is much more prominent than the renormalization of the gauge coupling and may explain the failure of (1a). Results from MC simulations of the $SU(2)$ lattice Higgs model for $\lambda = \infty$ indicate that this bound is relevant for the full model[6-7] and therefore provides a theoretical upper limit on the Higgs mass. The actual number is regularization dependent but this dependence may be weak.

2. THE FOUR COMPONENT Φ^4 MODEL

The observations of the preceding section motivate the regression from MC studies of the lattice Higgs model to a study of the Φ^4 model alone. The problem of the bound on the renormalized coupling has been attacked from various sides: expansion in the renormalized coupling taking into account the Gaussian scaling behavior,[14] MC simulations of the one-component[15-18] and the 4-component model[1,19,2]

The lattice action has the usual minimal form

$$
S = -\kappa \sum_{x \in \Lambda} \sum_{\mu=1}^{4} \left(\Phi_x^\alpha \Phi_{x+\mu}^\alpha + \Phi_x^\alpha \Phi_{x-\mu}^\alpha \right) \\
+ \lambda \sum_{x \in \Lambda} (\Phi_x^\alpha \Phi_x^\alpha - 1)^2 + \sum_{x \in \Lambda} \Phi_x^\alpha \Phi_x^\alpha + j \sum_{x \in \Lambda} \Phi_x^1,
$$

(2)

with the bare couplings κ, $\lambda \geq 0$. The fields Φ_x^α are real and summation over the index $\alpha = 1 \ldots 4$ is implied whenever it occurs pairwise. For the standard continuum normalization, where the coefficients of the kinetic and the mass term are $\frac{1}{2}$, the Φ fields should be rescaled to the fields $\varphi = \sqrt{2\kappa}\Phi$. All quantities will be given in dimensionless numbers (i.e., the lattice spacing $a = 1$).

Of course one may introduce further non-nearest neighbour two- and more-point interactions. Such actions will lead to different intrinsic lattice scales corresponding to different values of the cut-off.[20] There is a simple argument for the regularization dependence of the bound on the renormalized coupling. Let us assume that we have simulated the theory with the above action (2) for the value $\lambda = \infty$ (i.e., on the upper bound curve for this specific action) and generated an ensemble of configurations. Performing a block spin transformation on this ensembles does not change the long distance properties of the theory, it just changes the cut-off. Thus the blocked ensemble will have the same renormalized coupling (or, for the Higgs theory, the same ratio $\frac{m_H}{m_W}$) but the dimensionless value $\frac{1}{m_H}$ will be smaller by the scale factor of the transformation. The block-action corresponding to this block ensemble will be quite complicated, involving in principle infinitely many interaction terms; only for suitable block spin transformations one may hope that it is effectively represented by a few terms. The upper bound curve for a simulation based on this block action lies to the left of the curve for the simple action in (2). It is not universal, only the asymptotic behavior $\frac{1}{m_H} \to \infty$ is.

The updating method was a 2-hit Metropolis with more than 300000 MC-sweeps per point considered; every 4th configuration was evaluated. We worked for $j = 0$ at $\lambda = 0.05, 0.1, \infty$ and various values of κ in the broken phase, altogether 40 points. For $\lambda = \infty$ some results for non-zero j are presented elsewhere.[2] We studied ensembles of configurations for lattices of size L^4 ($L = 8, 12, 14$ and 16). The calculations were performed on vector computers Cyber 205 (SCRI and Universität Bochum), ETA^{10}(SCRI) and CRAY X-MP/48(KFA Jülich).

The $O(4)$ model is particular within the lattice simulations insofar it involves simultaneous occurence of massive and massless particle states. The three massless Goldstone modes make the calculation more problematic than for the one-component model. On a finite lattice the direction of symmetry breaking drifts through the group space, since, strictly speaking, there is no spontaneous symmetry breaking. We find that the direction of the mean value of Φ^α (taken on a configuration of size 8^4) covers the group space within a few 100 MC sweeps. This makes it difficult to disentangle the massive scalar and the light Goldstone modes. Introduction of an external field j gives additional mass to the Goldstone particles and stabilizes the direction of symmetry breaking.

We find that it is also possible to suppress the influence of the vacuum drift by selecting an appropriate field operator for the measurements. On a sufficiently large lattice for a given configuration the sum over fields

$$M^\alpha \equiv \frac{1}{L^4} \sum_x \Phi_x^\alpha \tag{3}$$

is an estimator for the direction of the spontaneously chosen vacuum on an infinite lattice. We introduce a field operator $\Phi_{s,x}$ by performing a global rotation such that the direction of M^α is rotated to the 1-axis in the group space separately for each configuration, i.e.,

$$\Phi_{s,x} = \frac{\Phi_x^\alpha M^\alpha}{\mid M \mid}. \tag{4}$$

In the infinite volume limit $\Phi_{s,x}$ corresponds to the massive scalar mode, its correlation function and expectation value give the scalar mass and field expectation value, respectively.

3. SOME RESULTS OF THE DIRECT MC STUDY

The expected scaling behavior is[21]

$$< \Phi > (t) \propto t^{1/2} (lnt)^{1/4}, m_s(t) \propto t^{1/2} (lnt)^{-1/4}, t = \frac{\kappa - \kappa_c}{\kappa_c}. \tag{5}$$

The scaling variable t may be replaced by other suitable variables like, e.g., the bare mass squared[17,18] or may be multiplied by arbitrary factors; only the behavior in the limit $t \to 0$ is universal. Due to the logarithm, a suitable choice of a multiplier may modify the "scaling behavior" substantially. For this reason a fit to the scaling behavior is not very conclusive unless the quality of the data allows to constrain this arbitrariness. In Fig. 2 we plot $< \varphi_s >^2$ versus κ for $\lambda = \infty$ and various lattice sizes.

We determine the mass of the massive scalar mode from the connected correlation functions of two operators. The first one is constructed from the field $\Phi_{s,x}$, as defined above,

$$O_1(\tau, \vec{p}) = \sum_{x \in \Lambda_\tau} e^{i\vec{x}\vec{p}} \Phi_{s,x}, \tag{6}$$

where Λ_τ denotes the 3-dimensional sublattice ("timeslice") at euclidean time τ. The second is a composite operator: O_2, the product $\Phi_x^\alpha \Phi_{x+\mu}^\alpha$ summed over the spatial directions, i.e., a non-local realization of Φ^2 (the latter is trivial for

164

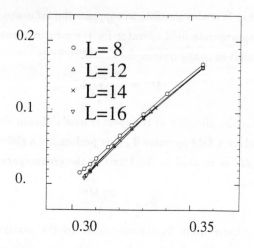

Figure 2. The field expectation value squared $< \varphi_s >^2$ as a function of κ for values of $L = 8, 12, 14, 16$ at $\lambda = \infty$: the leading scaling behavior should be linear in $(\kappa - \kappa_c)$. The slight deviation is compatible with the logarithmic contribution but not sufficiently significant.

$\lambda = \infty$). The mass is obtained by fitting the connected correlation function $C(T) = < O(0, \vec{p}) O(T, \vec{p}) >^c$ to the usual *cosh* behavior for massive state propagators (Fig. 3a). The mass values obtained from the propagators for $p = 0$ and $p > 0$ are in good agreement. A simultaneous fit of the data for $< \Phi_s >$ and m_s to the scaling behavior (5) leads to estimates for the "critical" values of κ (Fig. 4).

The measured propagator is that of the physical particle (on the lattice) and one has to take into account the renormalization of the field strength or, equivalently, of the kinetic term in the action corresponding to the p^2 term in the inverse propagator. Due to the Goldstone states there are infrared singularities in the propagators in the thermodynamic limit and therefore we determine the field renormalization constant Z from the momentum $p_0 = 2\pi/L$ propagator. The massive and the massless modes are members of one $O(4)$ multiplet, but in the broken phase their field renormalization may differ by finite contributions and their respective Z-factors do not have to be identical. This affects the relation between m_W and $< \varphi >$.[23,24] The value of Z_s for the massive scalar field decreases slowly with increasing mass but remains of order 0.9–1.0. The expectation value of the renormalized field is then $< \varphi_s^R > = < \varphi_s > / \sqrt{Z_s}$. The value for Z_G shows a

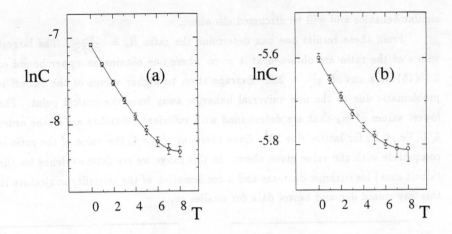

Figure 3. Propagators for the operators (a) O_1 and (b) O_3 for $L = 16, \lambda = \infty, \kappa = 0.305$; a fit to the cosh form gives (a) $m_s = 0.223$ and (b) $\mu = 0.084$.

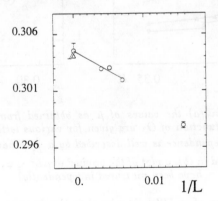

Figure 4. The values $\kappa_{PT,L}$ (circles) versus $\frac{1}{L^2}$ are consistent with the expected finite size scaling dependence $\kappa_{PT,L} - \kappa_\infty^c \propto L^{-\frac{1}{\nu}}$ for the Gaussian value $\nu = \frac{1}{2}$. The resulting value of $\kappa_\infty^c = 0.3045(7)$ comes from a fit to the values at $L = 12, 14, 16$ and is in good agreement with the value 0.3041 (triangle) from series expansions.[22]

similar behavior and will be discussed elsewhere.[2]

From these results one can determine the ratio $R_s = \frac{m_t}{<\varphi_s^R>}$. The largest values of the ratio are obtained at $\lambda = \infty$ where one obtains an upper bound of 2.50(15) at a cut-off $\frac{1}{m_s} = 2.5$. Extrapolation to higher values of the cut-off is problematic due to the non-universal behavior away from the critical point. The lowest values of m_s that are determined with sufficient reliability are of the order $4/L$, i.e., 0.25 for lattice size 16^4. Even there, at $\frac{1}{m_s} = 4$, the value of the ratio is compatible with the value given above. In this range we see little evidence for the (albeit slow) logarithmic decrease and a confirmation of the triviality conjecture in this way would demand better data for smaller m_s.

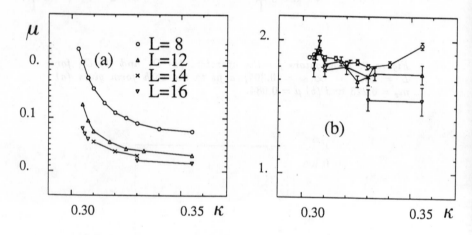

Figure 5. In (a) the values of μ as obtained from a fit to the correlation function of O_3 are given for various lattice sizes. The finite size dependence is well described by a $\frac{1}{L_2}$ behavior;[23] this is demonstrated in (b), a plot of the product of $\mu L^2 < \varphi_s >$ (the values μ and $< \varphi_s >$ have been measured independently).

The fluctuation of $< M^\alpha >$ is related to the decay properties of $O_3^\alpha(\tau, \vec{p}) = \sum_{x \in \Lambda_\tau} e^{i \vec{x} \vec{p}} \Phi_x^\alpha$. The corresponding correlation function (cf. Fig. 3b) is also well described by a *cosh* fit with an effective correlation length $\frac{1}{\mu}$, giving values for μ from 0.23 (close to the critical point) down to 0.02 (deeper in the broken phase). At first the identification of values that small seems to contradict the lore that one cannot observe "masses" below $O(2\pi/L)$ on finite lattices. A correlation length, which diverges at a second order PT arises from dynamical fluctuations of the system which cannot exceed the lattice size. As an example for a very different behavior take the lattice model of free bosons, the Gaussian model. It may be explicitly solved for any lattice size and arbitrarily small mass values. The propagator for $\vec{p} = 0$ states follows a precise *cosh* behavior. The lattice spacing follows a well-known analytic dependence and the critical κ has no finite size dependence.

The occurence of large correlation lengths $\frac{1}{\mu}$ is due to a mechanism different from that at a second order PT and has to do with the long distance ordering of the system at a first order phase transition; the corresponding correlation length is like a "coherence" or "persistence" length for co-existing phases.[25] Throughout the considered range μ shows a finite size dependence consistent with $1/L^2$ with deviations when approaching the critical point(cf. Fig. 5). Results on this correlation $< O_3^\alpha O_3^\alpha >$ may be used to obtain Z_G(cf. Ref. [2]).

4. CONCLUDING REMARKS

The proposed "rotated" field operator led to a consistent and convincing picture. However, one should check it by introduction of non-zero values of the external field j. Some results in this direction have been obtained[2] and a detailed study is in progress.

A direct MC simulation of the type presented here is a representative of one class of methods to study the mechanism of spontaneous symmetry breaking. Another class of methods is based on perturbation expansions. Both types of approach are very successful in the determination of physical observables like masses and each one provides an independent check on the quality of the other. A difficulty common to both approaches is to find convincing evidence whether the critical point is of Gaussian nature (cf. our problems to identify the logarithmic corrections) and whether this is the *only* fixed point of the Φ^4 model. This is due to the fact that one studies the critical behavior of the system close to but not at the PT. Universal behavior however may set in only very close to the critical point depending on the details of the system. Renormalization group methods seem to be better suited for this question of principle. The method of MC renormalization group works directly at the PT (or, on a finite lattice, on the point of maximal correlation length) and the block spin transformations keep the system at this point. Studies of the flow in coupling space give useful direct information on the fixed point structure and thus the possible universality class.[26] On the other hand, MCRG methods are not well suited to give results on masses or n-point functions. For a complete picture both approaches have their merits and should be pursued.

We may hope that eventually there will be rigorous results on the question of triviality in four dimensions. But even then MC methods as well as series expansions and other analytical techniques will be important tools to obtain detailed information on non-universal properties.

Acknowledgements

I want to thank Peter Hasenfratz and Peter Weisz for several helpful discussions. The support of the Computer Centers of Florida State University, Universität Bochum, and KFA Jülich is gratefully acknowledged. This work was supported by Jubiläumsfonds der Österreichischen Nationalbank.

REFERENCES

1. A. Hasenfratz, K. Jansen, C.B. Lang, T. Neuhaus and H. Yoneyama, Phys. Le, 199B, (1987) 531.
   K. Jansen, Nucl. Phys., B (Proc.Suppl.), (1988) 4.

2. A. Hasenfratz, K. Jansen, J. Jersák, C.B. Lang, T. Neuhaus and H. Yoneyama, in preparation.

3. J. Fröhlich, G. Morchio and F. Strocchi, Nucl. Phys., B190 [FS3], (1981) 553.

4. W. Langguth, I. Montvay and P. Weisz, Nucl. Phys., 277, (1986) 11.

5. H.G. Evertz, J. Jersák, C.B. Lang and T. Neuhaus, Phys. Lett., 171, (1986) 271.

6. W. Langguth and I. Montvay, Z. Phys, C36, (1987) 725.

7. A. Hasenfratz and T. Neuhaus, Nucl. Phys., B297 [FS21], (1988) 205.

8. H.G. Evertz, K. Jansen, J. Jersák, C.B. Lang and T. Neuhaus, Nucl. Phys., B285 [FS19], (1987) 590.

9. H.G. Evertz, RWTH Aachen, PhD Thesis, 1987.

10. I.-H. Lee and R. Shrock, Phys. Rev. Lett., 59, (1987) 14.
    I.-H. Lee and R. Shrock, Phys. Lett., B201, (1988) 497.
    I.-H. Lee and R. Shrock, and talks presented at this conference.
    A. Hasenfratz, talk presented at this conference.
    J. Polonyi and J. Shigemitsu, Ohio State Univ. Preprint DOE/ER/01545-403.
    D. Stephensen and A. Thornton, Edinburgh preprint 88/436.

11. cf. D.J.E. Callaway, See recent review, to be published in Phys. Rep.

12. P. Hasenfratz and J. Nager, Z. Phys., C37, (1988) 477 and Univ. Bern Preprint BUTP-87/18.

13. M. Lüscher and P. Weisz, Nucl. Phys., B290 [FS20], (1987) 27. ibid., Nucl. Phys., B295[FS21], (1988) 65.

14. C. Whitner, PhD thesis, Princeton University, 1984.

15. I. Montvay and P. Weisz, Nucl. Phys., B290 [FS20], (1987) 327.

16. K. Huang, E. Manousakis and J. Polonyi, Phys. Rev., D35, (1987) 3187.

17. J. Kuti and Y. Shen, Phys. Rev. Lett., 60, (1988) 85.

18. M.M. Tsypin, Lebedev Physical Institute Moscow preprint 280 1985. J. Kuti and Y. Shen, Nucl. Phys., B Proc. suppl. 4, (1988) 397.

19. K. Bitar, talk presented at this conference.

20. E. Brezin, J.C. Le Guillou and J. Zinn-Justin,Phase Transitions and Critical Phenomena, Vol. 6, eds. C. Domb and M.S. Green (Academic Press, 1976).

21. P. Weisz, talk presented at this conference.

22. H. Neuberger, preprints RU-32-87 1987 and RU-42-87 1987, and talk presented at this conference.

23. U.M. Heller and H. Neuberger, preprint NSF-ITP-88-13 1988.

24. M.E. Fisher and A.N. Berker, Phys. Rev., B26, (1982) 2507.

25. C.B. Lang, Phys.Lett., 155B, (1985) 399. ibid., Nucl. Phys., B265 [FS15], (1986) 630. D.J.E. Callaway and R. Petronzio, B240 [FS12], (1984) 577.

THE EFFECTS OF FERMIONS ON LATTICE GAUGE THEORIES WITH SCALARS

I-Hsiu Lee

Physics Department
Brookhaven National Laboratory
Upton, NY 11973

ABSTRACT

The effects of fermions on the gauge-Higgs systems are investigated in the context of an $SU(2)$ lattice gauge theory with scalar and fermion fields. The results from analytic studies and numerical simulations, with quenched and dynamical fermions, are presented. Some physical implications are discussed.

1. INTRODUCTION

One of our common goals in studying Higgs models on the lattice is to see if we can obtain some nonperturbative information about the electroweak theory. Most of the studies in this direction have been focused on the purely bosonic sector of the standard model. Recently, I have been concerned with another question, namely, what are the effects of fermions on these bosonic systems? In collaboration with J. Shigemitsu, R. Shrock and S. Aoki, I have investigated this question in various abelian and non-abelian models.[1-11] In this talk, I would like to present some of the results from our studies of this question in the context of an $SU(2)$ model with scalar and fermion fields.

1.1. The Model

Specifically, let us consider an $SU(2)$ gauge theory with a scalar field, ϕ, and four fermion fields, $\psi_f, f = 1, ...4$, all in the fundamental representation (i.e., with isospin $I_\phi = 1/2$, and $I_{\psi_f} = 1/2$), described by the Lagrangian density

$$\mathcal{L} = -\frac{1}{4}F_{\mu\nu}F^{\mu\nu} + \bar{\psi}_f i\gamma_\mu D^\mu \psi_f + (D_\mu\phi)^\dagger(D^\mu\phi) + \frac{1}{2}m^2\phi^\dagger\phi - \frac{1}{4}\lambda(\phi^\dagger\phi)^2 \quad (1)$$

The SU(2) indices are suppressed in the notation, and a sum over repeated indices is understood.

To study this model in a well-defined gauge-invariant manner, we construct the discretized path integral in usual notation

$$Z = \int \prod_{n,\mu} dU_{n,\mu} d\phi_n d\phi_n^\dagger d\chi_n d\bar{\chi}_n e^{-S} \quad (2)$$

where S is the discretized action of the model under consideration in Euclidean space-time, and

$$S = S_G + S_H + S_F \quad (3)$$

where

$$S_G = \beta_g \sum_{plaq.} [1 - P] \quad (4)$$

$$S_H = 2\beta_h \sum_{n,\mu} Re(\phi_n^\dagger\phi_n - \phi_n^\dagger U_{n,\mu}\phi_{n+e_\mu}) + \frac{1}{4}\lambda\beta_h^2 \sum_n (\phi_n^\dagger\phi_n - 1)^2 \quad (5)$$

$$S_F = \frac{1}{2} \sum_{n,\mu} \eta_{n,\mu}(\bar{\chi}_n U_{n,\mu}\chi_{n+e_\mu} - \bar{\chi}_{n+e_\mu}U_{n,\mu}^\dagger\chi_n) \quad (6)$$

with $\beta_g = 4/g^2$, $\beta_h = m^2/\lambda$, and $P = (1/2)Tr(U_{plaq.})$. The site n lies in the d-dimensional hypercubic lattice. Unless otherwise specified, $d = 4$ and the units are chosen so that the lattice spacing $a \equiv 1$. We use staggered fermions, which are advantageous for studies of chiral symmetry. The $\eta_{n,\mu}$ are factors from the Dirac matrices and are given by 1 for $\mu = 1$ and $(-1)^{n_1+\dots+n_{\mu-1}}$ for $2 \leq \mu \leq d$.

Let me begin by reviewing the phase diagram of this model in the absence of fermions.

1.2. Phase Diagram in the Absence of Fermions

The purely bosonic part of the model with the lattice action $S = S_G + S_H$, has been studied extensively. Its phase diagram is well known.[12,13] In particular, the

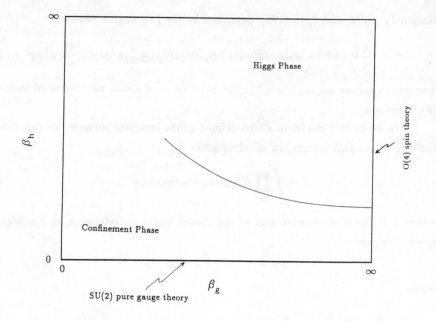

Figure 1. Schematic phase diagram in the $\beta_g - \beta_h$ plane for the SU(2) gauge theory with a Higgs field in the fundamental representation.

phase diagram in the $\beta_g - \beta_h$ plane for a large range of λ is known to be qualitatively similar to that for the limiting case $\lambda \to \infty$, which is shown in Fig. 1. For simplicity, let us focus our discussions on the $\lambda \to \infty$ limit here. In this limit, the scalar fields satisfy $\phi_n^\dagger \phi_n = 1$, and

$$S_H = -2\beta_h \sum_{n,\mu} Re(\phi_n^\dagger U_{n,\mu} \phi_{n+e_\mu}) \tag{7}$$

The properties of several limiting cases of this model are easy to establish from Eqs. (2), (4) and (7). For $\beta_h = 0$, the model reduces to a pure SU(2) gauge theory. On the opposite side of the phase diagram as $\beta_h \to \infty$, the model becomes trivial. For $\beta_g = \infty$, the gauge degrees of freedom are frozen, and the model becomes an O(4) spin model, which is known to have a second-order phase transition at a finite β_h where the global O(4) symmetry is spontaneously broken to O(3). Numerical simulations of this model suggest that the O(4) transition is the endpoint of a line of transitions which extends into the interior of the phase diagram. However, this line of transitions cannot extend all the way to $\beta_g = 0$, but must end in the interior of the phase diagram. The reason is that, at $\beta_g = 0$, the model is exactly solvable, and the free energy is analytic for all values of β_h, which implies that there can be no phase transition for $\beta_g = 0$. Furthermore, since a small-β_g expansion has a finite radius of convergence, the qualitative features at $\beta_g = 0$ still hold for a finite range of small $\beta_g > 0$ where there is no phase transition either.

Most notable of this phase diagram is the fact that the confinement and the Higgs phases are analytically connected. This has inspired ideas such as complementarity[14] and the Abbott-Farhi model.[15]

Having reviewed the situations in the absence of fermions, we may now ask: what happens when fermions are included in the model?

2. EFFECTS OF FERMIONS

2.1. Studying a Vectorlike Theory with Quenched Approximation

There have been many studies which suggest that in $SU(N)$ gauge theories, such as QCD, where the fermions couple to the gauge field in a vectorial manner, chiral symmetry is spontaneously broken for nonzero couplings. This implies that, if we include in our model fermions which couple to the gauge field vectorially, we would find the chiral condensate being nonzero for finite values of β_g along the line $\beta_h = 0$. Since a small β_h-expansion has a finite radius of convergence, the same is true for a finite range of small but positive β_h. On the other hand, as was mentioned in section 1.2, for $\beta_h = \infty$, the fermions become free. There exists no interaction to trigger the spontaneous breakdown of chiral symmetry. The question then is: how does the chiral condensate vanish as a function of β_h? Does it go to zero only when β_h goes to infinity, or is there a finite β_h at which the condensate vanishes non-analytically?

This question was first addressed in Ref. [1], where a numerical simulation of the model defined by Eqs.(2)-(4) and (6)-(7) was performed using the quenched approximation. We chose to work at $\beta_g = 0.5$ where all observables are analytic functions of β_h in the absence of fermions. [The endpoint of the line of transitions in the interior of the phase diagram Fig. 1 occurs at approximately[16] $(\beta_g, \beta_h) \approx (1.6, 0.65)$.] We measured the chiral condensate at several values of β_h; the results are

174

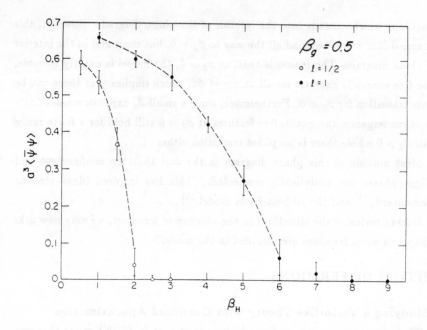

Figure 2. Chiral condensate as a function of β_h at $\beta_g = 0.5$ (measured in the quenched approximation).

presented in Fig. 2. The data show evidence for the existence of a chiral transition at a finite β_h. We also found that, in the small-β_h region where the chiral condensate is nonzero, a massless pion appears in the spectrum, while in the large-β_h region where the chiral condensate vanishes, massless fermions exist. Therefore, the simulation with the quenched approximation suggests that a new transition associated with the fermions exists.

2.2. Relevance: Chiral Theory vs. Vectorlike Theory

At first sight, the new transition we have found may seem to be only of theoretical interest, since we had considered a vectorlike theory where fermions couple to the gauge field vectorially, while the SU(2) part of the standard model is a chiral theory with the left-handed and the right-handed fermions transforming differently under gauge transformations. However, since the representations of SU(2) are real, a chiral theory with $2N_f$ fermion doublets, for instance, can be written equivalently as a vectorlike theory with N_f doublets.[17,11] In the model which we are considering, $N_f = 4$. Therefore, neglecting the Yukawa couplings, we can consider our model as one which describes the $SU(2)_L$ part of the standard model with two generations.[1] Then the results of the quenched simulation show that this latter model has a phase transition associated with the spontaneous breaking of chiral symmetry.

Now that we have seen that the new transition we found carries potentially important nonperturbative information about the standard model, the next question is: how reliable is the quenched approximation? Is there a way to demonstrate the existence of the chiral transition analytically?

2.3. Analytic Studies

Recall that, in the absence of fermions, we know that the confinement and the Higgs phases are analytically connected to each other, since (i) we can calculate the free energy at $\beta_g = 0$ and find that it is analytic, and (ii) the properties at β_g still hold for a finite strip adjacent to the line $\beta_g = 0$. Conversely, if we can show that, in the presence of fermions, a chiral transition exists at $\beta_g = 0$, then the same argument implies that the chiral transition still exists for a finite strip at small β_g. However, as was noted in Ref. [3], this line of chiral transitions cannot stop in the interior of the phase diagram, since it separates a region where the chiral condensate is nonzero from another where the condensate vanishes identically, and these two regions cannot be analytically connected to each other. Therefore, if there is a chiral transition at $\beta_g = 0$, there must exist a line of chiral transitions which completely separates the phase diagram into two disjoint regions. The main question now is whether one can show the existence of a chiral transition at $\beta_g = 0$.

We have carried out a series of analytic studies[2,3,5−8,10,11] of various models in the strong gauge coupling limit (i.e. $\beta_g = 0$). We studied both abelian and non-abelian models with scalar and fermion fields, using a mean-field type of approximation. One of the main results of these studies is that the existence of chiral transition is a very general feature of lattice gauge theories with scalar and

---

[1] The gauge singlet right-handed fermions can be included in the model. However, without the Yukawa couplings, they have no effect since they completely decouple from the rest of the system.

fermion fields. Some of these works are discussed in R. Shrock's contribution at this workshop. I will only mention the results for the model which we are considering here, namely, the SU(2) gauge theory with a scalar field and a staggered fermion field, both in the fundamental representation. With the mean field approximation, we found that[6] the chiral condensate decreases monotonically and continuously as β_h increases from zero to $\beta_h \approx 2.76$, beyond which the condensate vanishes identically. This implies that there is a second-order chiral transition at $\beta_g = 0$, and the critical point is at $\beta_{h,c} \approx 2.76$.

This analytic result can be compared with the data from our quenched simulation.[6] In Fig. 3, the chiral condensate $M \equiv\, <\bar\chi\chi>$ is plotted as a function of β_h. We see that the condensate decreases monotonically, and the data suggests that a chiral transition occurs at approximately $\beta_{h,c} \approx 2.7$. This is in good agreement with the analytic result.

As I mentioned before, although our analytic studies were carried out at the strong gauge coupling limit, the result is actually more general. We expect that the chiral transition found at $\beta_g = 0$ extends into the interior of the phase diagram all the way until it reaches one of the two boundaries, $\beta_h = 0$ or $\beta_g = \infty$. Furthermore, it is worth noting that the chiral transition is determined to be of second-order at $\beta_g = 0$. One expects from a small-β_g expansion that the transition will remain second-order for at least a finite range of β_g.

2.4. Simulation with Dynamical Fermions

To get a better feeling of how fermions affect the bosonic system, especially for nonzero β_g, we performed a simulation with dynamical fermions.[10] The results of our simulation are presented in the following. Readers are referred to Ref. [10] for the technical details.

We first study the $\beta_g = 0$ case, since we have some analytic results here to compare with the simulation. Three quantities are measured, the chiral condensate $<\bar\chi\chi>$, the plaquette $<P>$ and $<L> \equiv\, <\sum_{n,\mu} Re(\phi_n^\dagger U_{n,\mu}\phi_{n+e_\mu})/N_\ell>$, where N_ℓ is the number of links in the lattice. To measure $M \equiv\, <\bar\chi\chi>$, we have to add a small bare mass as the source to calculate $M_m \equiv\, <\bar\chi\chi>_m$, and then extrapolate to zero bare mass limit. The data for M_m at $\beta_g = 0$ are shown in Fig. 4a for bare mass $m = 0.1$ and 0.05. The extrapolation of these data to the zero bare mass limit is shown in Fig. 4b. It shows that the chiral condensate decreases as β_h increases,

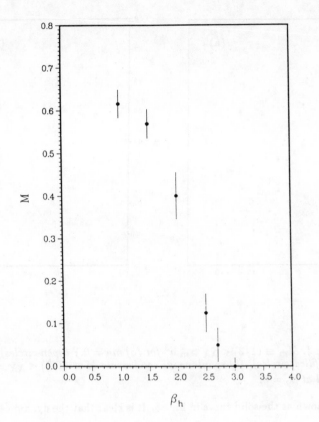

Figure 3. Chiral condensate as a function of β_h at $\beta_g = 0$ (measured in the quenched approximation).

and vanishes for sufficiently large β_h. The critical point of the chiral transition is estimated to be $\beta_h = 2.5 \pm 0.3$, which is close to the results from the analytic study (with a mean-field approximation) and the quenched simulation.

The data for $< L >$ with $m = 0.1$ and 0.05 are plotted in Fig. 5 as open and filled circles, respectively. They are so close to each other that the open circles can hardly be seen. To see the effects of fermions on $< L >$, recall that, in the absence of fermions, $< L >$ can be calculated exactly:

$$< L >_{bos.} (\beta_g = 0) = I_2(2\beta_h)/I_1(2\beta_h). \qquad (8)$$

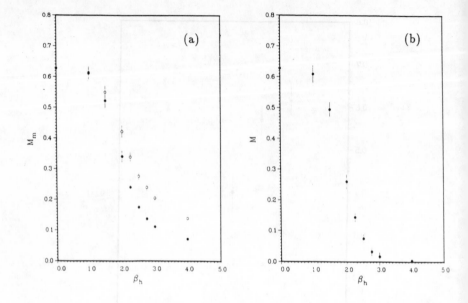

Figure 4. $M_m \equiv (1/2) < \bar{\chi}\chi >_m a^3$ *for (a)* $ma = 0.1$ *(open circles),* 0.05 *(filled circles); (b) spontaneous condensate* $M \equiv (1/2) < \bar{\chi}\chi > a^3$, *all at* $\beta_g = 0$.

$< L >_{bos.}$ is shown as the solid curve in Fig. 5. It is clear that the dynamical fermion data of $< L >$ are very close to $< L >_{bos.}$, indicating that the fermion effects on $< L >$ are negligible for $\beta_g = 0$. Although we do not see any obvious singularity in $< L >$ from the data, our analytic result shows that $< L >$ has a non-divergent singularity at the critical point of the chiral transition. Such singularity is difficult to observe numerically, however.

The data for $< P >$ are also plotted in Fig. 5 as open and filled squares for $m = 0.1$ and 0.05, respectively. In the absence of fermions,

$$< P >_{bos.} (\beta_g = 0) = [I_2(2\beta_h)/I_1(2\beta_h)]^4. \qquad (9)$$

In Fig. 5, $< P >_{bos.}$ is plotted as the dashed curve. The effects of fermions on $< P >$ are more appreciable. This may be due to the fact that, in our model, fermions couple to the gauge fields directly while they interact with the scalars only through gauge fields.

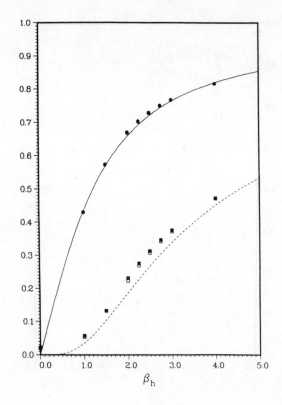

Figure 5. $< L >$ for $ma = 0.1$ (open circles), 0.05 (filled circles);
$< P >$ for $ma = 0.1$ (open squares), 0.05 (filled squares), all
at $\beta_g = 0$. Solid and dashed curves represent $< L >_{bos.}$ and
$< P >_{bos.}$, respectively, at $\beta_g = 0$.

Next, we study the interior of the phase diagram. As was mentioned earlier,
it is expected that the qualitative features established at $\beta_g = 0$ remain the same
for a finite strip adjacent to $\beta_g = 0$. This expectation is borne out by our data for
$\beta_g = 0.5$. In Fig. 6, the chiral condensate at $\beta_g = 0.5$ is shown. It behaves very
similarly to the $\beta_g = 0$ case (except that the critical point of the chiral transition
now moves to a lower value of β_h). The same is true of other quantities we measured.
According to our data, the chiral transitions at $\beta_g = 0$ and 0.5 are both consistent

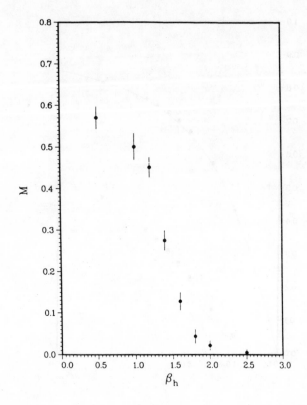

Figure 6. $M \equiv (1/2) < \bar{\chi}\chi > a^3$ at $\beta_g = 0.5$.

with being of second-order, in agreement with the result of our analytical study.

So far, we have been studying the region of the phase diagram where there is no phase transition in the absence of fermions. To see how the chiral transition and the confinement-Higgs transition affect each other, we study the case $\beta_g = 1.9$. In Fig. 7, M_m at $\beta_g = 1.9$ is plotted for $m = 0.1$ and 0.05. If we extrapolate the data to the zero bare mass limit, we find that the chiral transition occurs at $\beta_h \approx 0.4$. To see whether there is a separate confinement-Higgs transition, we plot the data for $< L >$ at $\beta_g = 1.9$ in Fig. 8 as the open and filled circles for $m = 0.1$ and

0.05, respectively. The data show no evidence for other transition than the one at $\beta_h \approx 0.4$. In contrast, the same simulation for the purely bosonic system shows that the confinement-Higgs transition occurs at $\beta_h \approx 0.52$. Furthermore, for a given β_h, the data for $< L >$ with dynamical fermions is always greater than that in the purely bosonic system. It implies that fermion effects make the system more ordered. For comparison, we include $< L >$ at $\beta_g = \infty$ (open squares) and $< L >_{bos.}$ at $\beta_g = 0$ (solid curve) in Fig. 8. Since the fermion effects are negligible at $\beta_g = 0$, as we mentioned before, the solid curve can be regarded as a good approximation of $< L > (\beta_g = 0)$. It appears that the behavior of $< L >$ at $\beta_g = 1.9$ is more similar to that at $\beta_g = \infty$ than at $\beta_g = 0$.

From our simulation with dynamical fermions, we obtain the phase diagram for our model in the presence of fermions, as shown in Fig. 9.

3. CONCLUSION

In summary, we discussed the effects of fermions in a vectorlike $SU(2)$ gauge theory with a scalar field in the fundamental representation. We found that, in the presence of fermions, (i) there exists a new transition associated with chiral symmetry breaking/restoration, which completely separates the phase diagram into two disjoint regions, (ii) the transition is second-order for, at least, a finite range of β_g, and (iii) for fermions in the fundamental representation, there is only one transition at $\beta_g = 1.9$ (i.e. there is no separate confinement-Higgs transition).

What physical implications can we draw from here? First, interpreting our model as the $SU(2)_L$ part of the standard model with two generations (cf. section 2.2), our results imply that a chiral transition exists in the latter lattice model. This is obviously a nonperturbative effect. Secondly, since the chiral transition is continuous for at least a finite range of β_g, it opens up new possibilities for the continuum limit of the lattice theory to be taken. It remains to be seen whether a physically interesting continuum limit can be defined along the chiral transition phase boundary. Finally, I would like to briefly comment on the implications of our results for the Strongly Coupled Standard Model (SCSM).[18]

The lattice model which we consider in this talk can also be regarded as the discretized version of the SCSM. In SCSM, it is assumed that (i) the $SU(2)$ sector of the standard model remains confined so that the physical spectrum only consists of gauge singlet bound states and their excited states. This assumption

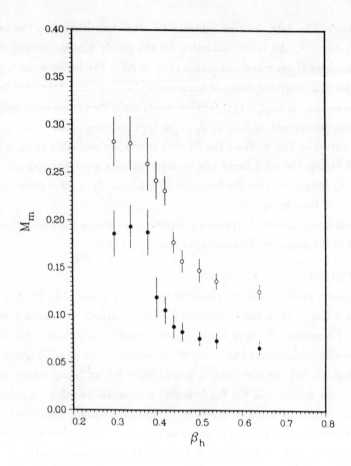

Figure 7. $M_m \equiv (1/2) < \bar{\chi}\chi >_m a^3$ at $\beta_g = 1.9$ for $ma = 0.1$ *(open circles),* 0.05 *(filled circles).*

is naturally satisfied in the small-β_h region of the phase diagram (Fig. 9), where the spectrum is similar to that of QCD.[1,19] Furthermore, to ensure the existence of light fermions in the spectrum, the SCSM assumes that (ii) chiral symmetry is not spontaneously broken. However, Fig. 9 indicates that this assumption is not satisfied in the small-β_h region. In fact, to get a spectrum with light fermions, it is necessary to approach the continuum limit of the lattice model from within the chirally symmetric phase. The only possibility for both assumptions (i) and (ii) to hold is for the chirally symmetric phase of the phase diagram, Fig. 9, to

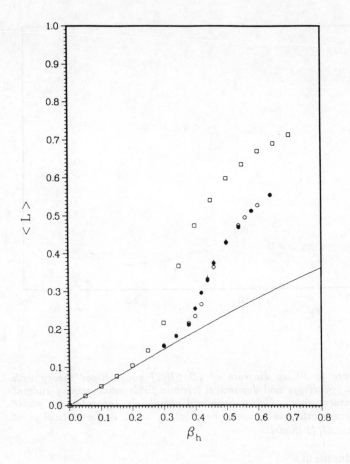

Figure 8. $< L >$ *at* $\beta_g = 1.9$ *for* $ma = 0.1$ *(open circles), 0.05 (filled circles). Solid curve is* $< L >_{bos.}$ *($\beta_g = 0$) and data denoted by open squares is* $< L > (\beta_g = \infty)$.

have the properties required by assumption (i). Although the evidence from Refs. [1] and [19] is not strong enough to rule out this possibility, it does suggest that the spectrum in the chirally symmetric phase is rather different from that in an SU(2) QCD-like theory. Moreover, the presence of the chiral boundary separating the chirally symmetric and broken phases removes the basis for the principle of complementarity,[14] which helped to motivate the strongly coupled standard model.

184

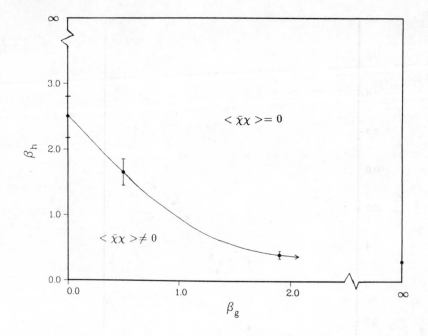

Figure 9. Phase diagram of 4D SU(2) gauge-Higgs theory with $I = 1/2$ Higgs and dynamical fermion fields summarizing current measurements. Chiral boundary completely separates the phase diagram into two phases. Point at $\beta_g = \infty$ is the critical point of the O(4) model.

Acknowledgements

This research was supported in part by the U.S. Department of Energy under contracts DE-AC02-76CH00016.

REFERENCES

1. I-H. Lee and J. Shigemitsu, Phys. Lett. **178B** (1986) 93.
2. I-H. Lee and R. E. Shrock, Phys. Rev. Lett. **59** (1987) 14.
3. I-H. Lee and R. E. Shrock, Nucl. Phys. **B290** (1987) 275.
4. I-H. Lee and R. E. Shrock, Phys. Lett. **196B** (1987) 82.
5. I-H. Lee and R. E. Shrock, Phys. Lett. **199B** (1987) 541.
6. I-H. Lee and R. E. Shrock, Phys. Lett. **201B** (1988) 497.
7. I-H. Lee, in Proceedings of the International Symposium on Field Theory on the Lattice, Seillac, Nucl. Phys. B (Proc. Suppl.) 4 (1988), 373.

8. R. E. Shrock, in Proceedings of the International Symposium on Field Theory on the Lattice, Seillac, Nucl. Phys. B (Proc. Suppl.) 4 (1988), 438.

9. S. Aoki, I-H. Lee, and R. E. Shrock, Phys. Lett. B, in press.

10. I-H. Lee and R. E. Shrock, BNL-Stony Brook preprint BNL-41287/ITP-SB-88-7

11. I-H. Lee and R. E. Shrock, BNL-Stony Brook preprint BNL-ITP-SB-88-27

12. C. B. Lang, C. Rebbi and M. Virasoro, Phys. Lett. **104B** (1981) 294.

13. H. Kühnelt, C. B. Lang and G. Vones, Nucl. Phys. **B230** [FS10] (1984) 16, and references therein.

14. S. Dimopoulos, S. Raby and L. Susskind, Nucl. Phys. **B173** (1980) 208.

15. L. F. Abbott and E. Farhi, Phys. Lett. **101B** (1981) 69; Nucl. Phys. **B189** (1981) 547.

16. I. Montvay, Phys. Lett. **150B** (1985) 441.

17. H. Georgi, as cited in Ref. 18.

18. M. Claudson, E. Farhi, and R. L. Jaffe, Phys. Rev. **D34** (1986) 873.

19. A. De and J. Shigemitsu, Nucl. Phys. B, in press.

HOW FAR IS N=4 FROM N=∞ ?

Lee Lin* with Julius Kuti and Yue Shen

Department of Physics
University of California at San Diego
La Jolla, CA 92093, USA

ABSTRACT

It is shown that results from the large N expansion do not provide useful quantitative guidance for non-perturbative questions in the physically relevant $O(4)$ scalar field theory with quartic self-interaction. We expect similar difficulties with the large N expansion in other applications, like in estimating a triviality upper bound on the mass of a heavy top quark in the standard electroweak model. The non-perturbative upper bound on the mass of the Higgs particle is better estimated by renormalization group improved summation of Feynman diagrams for $N = 4$. Renormalized perturbation theory accounts for lattice simulation of Euclidean Green's functions in the scaling region of the $O(4)$ model on both sides of the critical line. We find no evidence for a strongly interacting Higgs sector for dimensionless Higgs correlation lengths $\xi > 2$ on the lattice. For $\xi > 2$ we will show that the width of the Higgs particle is less than twenty percent of the mass and this relatively narrow resonance can be studied very effectively with conventional lattice techniques.

1. INTRODUCTION

In our Monte Carlo lattice simulation of the $O(4)$ model we have found the scaling region on both sides of the critical line.[1,2] In the scaling region, where one has an effective theory with non-vanishing interaction at finite lattice cut-off, physical results will not be sensitive to the presence of a cut-off, to a very good accuracy. The small scaling violation terms, however, grow rapidly with increasingly heavy Higgs mass. Physical observables become dominated by regularization artifacts as we leave

---

* speaker at the conference

the scaling region. As the mass of the Higgs particle is tuned to approximately three times the vacuum expectation value of the Higgs field, or larger, we are driven away from the scaling region in the O(4) model. This observation sets an upper bound on the mass of the Higgs particle as it was discussed in earlier talks at this conference.

We have shown evidence from our Monte Carlo simulations that physical observables are reasonably well approximated by renormalized perturbation theory in the scaling region.[1,2] In other words, the Higgs sector breaks down and becomes dominated by cut-off artifacts only for strong Higgs self-coupling, or equivalently, for large Higgs masses above the 600 GeV region. This scenario is now strongly supported by using lattice regularization in the calculations. Although we do not expect significant changes when a different type of cut-off is applied, the regularization scheme dependence of the results will require further studies.[3]

The list which follows is a brief summary of some of the results within the framework of renormalized perturbation theory which are supported by our Monte Carlo simulation in the O(4) model:[1,2]

1. The quantitative behavior of the finite lattice propagator in momentum space and its similarity to a free propagator is well understood in perturbation theory.

2. The infrared behavior generated by massless Goldstone excitations is well-controlled by the finite volume regulator diagrammatically.

3. Renormalization group improved loop expansion for the constraint effective potential agrees well with our Monte Carlo results.

4. A qualitative estimate of the Higgs bound based on the renormalization group improved perturbation theory relation between bare and renormalized coupling works surprisingly well when compared with our numerical results.

Recently, it was suggested that the large N expansion of the O(N) model leads to a different scenario.[4] Taking large N results literally, one finds a Higgs mass bound which is higher than it was found in our Monte Carlo calculations. In addition, the heavy Higgs particle with a mass close to 1 TeV becomes a very broad resonance in the large N expansion. This implies then a strongly interacting Higgs sector with a massive Higgs particle which appears as a very broad resonance (complex pole) in the Higgs propagator. Lattice techniques we used in our work earlier would require then very significant modifications before useful results could be extracted.

We would like to demonstrate here that although the large N results are mathematically correct, they are not applicable to the $O(4)$ model in a quantitative way. First, we will reproduce the large N expansion for completeness, and for comparison purposes.

2. LARGE N EXPANSION IN THE O(N) MODEL

The O(N) Lagrangian is defined in continuum notation as

$$\mathcal{L} = \frac{1}{2} \sum_{i=1}^{N} (\partial_\mu \Phi_i)(\partial_\mu \Phi_i) - \frac{1}{2} m^2 \sum_{i=1}^{N} \Phi_i^2 + \lambda \Big(\sum_{i=1}^{N} \Phi_i^2 \Big)^2 + counterterms , \quad (1)$$

where m^2 is a positive mass parameter in the phase with spontaneously broken symmetry, and λ designates the renormalized quartic Higgs self-coupling at some momentum scale μ. In the broken symmetry phase we shift the Nth component of the O(N) field Φ_i by the vacuum expectation value v, and we define new fields by the relations $\Phi_i = \xi_i$, for $i = 1, 2, ..., N-1$, and $\Phi_N = h + v$. After the field shift, the $O(4)$ Lagrangian becomes

$$\mathcal{L} = \frac{1}{2} \partial_\mu h \partial_\mu h + \frac{1}{2}(2m^2)h^2 + \frac{1}{2} \sum_{i=1}^{N-1} \partial_\mu \xi_i \partial_\mu \xi_i + \lambda h^4 + 4\lambda v h^3$$

$$+ 4\lambda v h \sum_{i=1}^{N-1} \xi_i^2 + 2\lambda h^2 \sum_{i=1}^{N-1} \xi_i^2 + \lambda \sum_{i=1}^{N-1} \xi_i^4 + \lambda \sum_{\substack{i \neq j \\ i,j=1}}^{N-1} \xi_i^2 \xi_j^2$$

$$+ counterterms . \quad (2)$$

Eq. (2) describes a heavy Higgs particle h with renormalized mass $m_H = \sqrt{2}m$, in addition to $N-1$ massless Goldstone particles with fields ξ_i.

Figure 1 shows the leading diagrams which contribute in the symmetric phase to the four-point Euclidean Green's function in the large N limit. The bubble diagrams of Figure 1 form a geometric series which can be summed to obtain

$$\lambda(\mu) = \frac{\lambda(\Lambda)}{1 + \frac{N}{2\pi^2}\lambda(\Lambda) \cdot \ln\left(\frac{\Lambda}{\mu}\right)} , \quad (3)$$

where the renormalized coupling constant $\lambda(\mu)$ is defined at some Euclidean subtraction point μ of the four-point function ($\lambda(\Lambda)$ is the bare coupling constant defined at the cut-off scale Λ). One reason for the non-zero Euclidean scale μ in Eq. (3) is that the renormalization properties of Euclidean Green's functions cannot be defined at zero momentum where infrared singularities are present with massless Goldstone particles in the broken symmetry phase.

$+$ 2 other channels

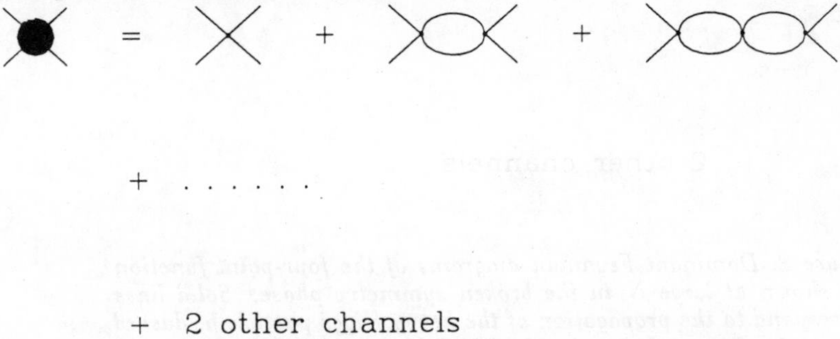

Figure 1. Dominant Feynman diagrams of the four-point function at large N are shown in the symmetric phase. Solid lines designate propagating massive fields with N components. Other diagrams for fixed λ are smaller by some power counting of N.

In the broken symmetry phase we find that bubble diagrams with Goldstone particles dominate, as depicted in Figure 2. Summing the geometric series of Figure 2 we get the same formula as in Eq. (3). For fixed cut-off Λ the large N behavior of the renormalized coupling is

$$\lambda(\mu) \approx \frac{2\pi^2}{N \cdot \ln(\Lambda/\mu)} \ . \tag{4}$$

Using similar large N summation techniques, we can calculate the renormalized Higgs particle propagator in the broken symmetry phase. Figure 3 shows the dominant Feynman diagrams in the large N limit. The large N geometric summation of diagrams in Figure 3 gives the momentum space Higgs propagator (defined as the two-point function of the h field) with the form

$$\tilde{G}^{(2)}(p) = \frac{1 + 4\lambda(\mu)N \cdot I}{p^2 \left[1 + 4\lambda(\mu)N \cdot I\right] + 2m^2} \ , \tag{5}$$

where the tree level relation $m^2 = 4\lambda v^2$ will be used in our numerical calculations. The integral I in Eq. (5) has a subtraction to make it convergent, with the result

$$I = \int \frac{d^4k}{(2\pi)^4} \frac{1}{k^2(k-p)^2} + counterterm$$

+ 2 other channels

Figure 2. Dominant Feynman diagrams of the four-point function are shown at large N in the broken symmetry phase. Solid lines correspond to the propagation of the heavy Higgs particle h, dashed propagator lines are associated with Goldstone particles.

$$= \frac{1}{16\pi^2} \ln(\frac{\mu^2}{p^2}) . \tag{6}$$

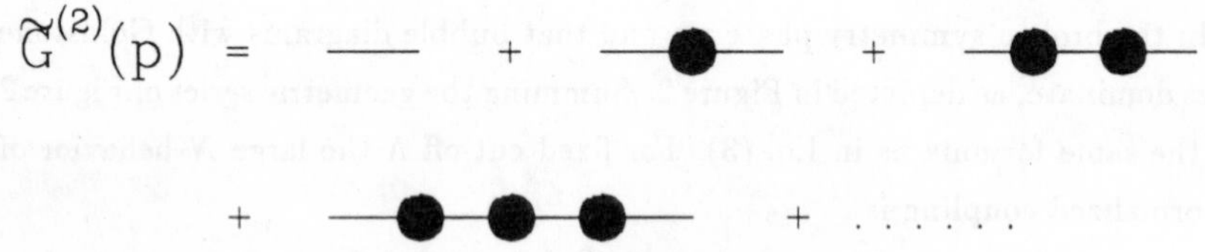

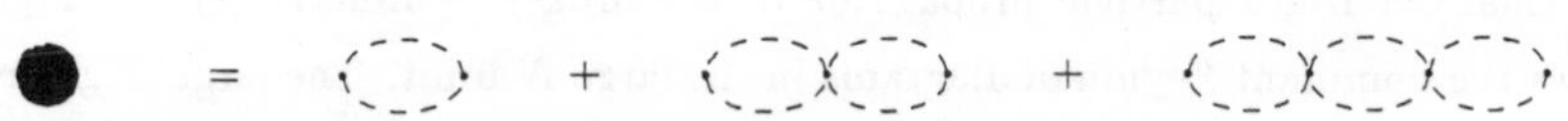

Figure 3. Dominant Feynman diagrams of the Higgs particle propagator at large N. Other diagrams for fixed λ are smaller again by some power of N.

We will now show that the large N results, which are identical to the results of an earlier work,[4] lead to a very broad Higgs resonance with strong self-interaction in the heavy Higgs mass limit.

3. BROAD HIGGS RESONANCE AT LARGE N

The two-point function, which was just calculated in large N expansion, exhibits the expected analytic properties. The function is analytic at Euclidean momenta, for $p^2 > 0$. It has a branch cut starting at zero p^2 due to the presence of Goldstone particles as massless intermediate states. The heavy Higgs particle corresponds to a complex pole in the Minkowski region, below the branch cut, on the second sheet of the complex p^2 plane. The physical properties of the Higgs particle are determined by the location of the complex pole. With the notation

$$\frac{1}{\tilde{G}^{(2)}(-p^2 = S_H)} = 0 , \qquad (7)$$

where $S_H = (m_H - i\frac{1}{2}\Gamma_H)^2$ is the location of the pole in the complex plane, we find

$$S_H = \frac{2m^2}{1 + \frac{N\lambda}{4\pi^2} \ln(\frac{\mu^2}{-S_H})} . \qquad (8)$$

In Eq. (8) m_H designates the renormalized mass of the Higgs particle and Γ_H, which is proportional to the imaginary part of the complex pole, designates the width of the Higgs particle. To solve for S_H we set $\mu^2 = v^2$, where $v \approx 250\ GeV$. The pole position in the large N limit is depicted in Figure 4 where the Higgs correlation length ξ is calculated for each point.

In the large N expansion Figure 4 indicates that the imaginary part of the complex Higgs pole is comparable, or larger than the real part when the Higgs correlation length ξ becomes comparable to one in lattice spacing units. As shown in Figure 4, the width Γ_H is comparable to the heavy Higgs mass for $\xi \approx 1$ indicating strong coupling regime in the Higgs sector. To get to the weakly interacting perturbative regime one has to choose a correlation length as large as $\xi \approx 80$. This result is very different from what we have found in the simulation of the O(4) model where perturbation theory was only breaking down for $\xi \approx 1$. The resolution of this puzzle is that the large N expansion is not a good approximation to the O(4) model.

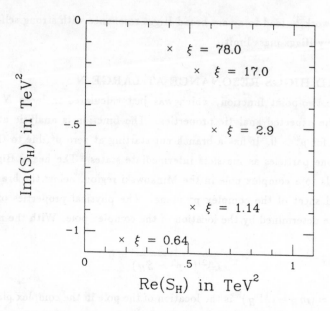

Figure 4. The position of the complex Higgs pole in the propagator is displayed in large N approximation as a function of the dimensionless Higgs correlation length ξ which is related to the lattice momentum cut-off Λ by the relation $\pi \cdot \xi = \Lambda/m_H$. The Higgs correlation length is shown for each calculated point.

4. N=4 IS FAR FROM INFINITY

We can solve the renormalization group equation for the running coupling constant with the one-loop beta function for arbitrary N. From the solution we find

$$\frac{1}{\lambda(\mu)} = \frac{1}{\lambda(\Lambda)} + \frac{N+8}{2\pi^2} \cdot \ln(\Lambda/\mu) . \tag{9}$$

When we approach the Landau pole we get

$$\lambda(\mu) = \frac{2\pi^2}{(N+8) \cdot \ln(\Lambda/\mu)} , \tag{10}$$

in the limit $\lambda(\Lambda) \to \infty$. Eq. (10) is the upper bound on the renormalized Higgs coupling $\lambda(\mu)$ in renormalization group improved perturbation theory. We have seen earlier at this conference[1] that the qualitative estimate of Eq. (10) is surprisingly close to the Monte Carlo results of the O(4) model.

We can also calculate the momentum space propagator in renormalization group improved perturbation theory for arbitrary N using the one-loop beta function for the running coupling constant. We find

$$\tilde{G}^{(2)}(p) = \frac{1 + 4\lambda(\mu)(N-1) \cdot I}{p^2 \left[1 + 4\lambda(\mu)(N-1) \cdot I\right] + 2m^2} \; , \tag{11}$$

where

$$I = \frac{1}{16\pi^2} \; \ln(\frac{\mu^2}{p^2}) \; . \tag{12}$$

We note that the factor N in the denominator of the large N formula in Eq. (5) is replaced by $N - 1$.

We can go through the same exercise as before and solve for the complex pole of the Higgs propagator using Eqs. (10-12). The results are presented in Figure 5 for infinite bare coupling.

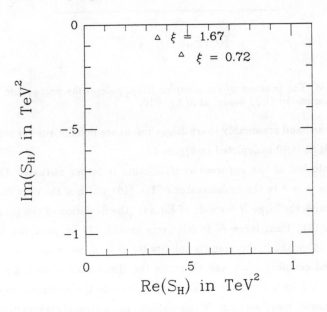

Figure 5. The position of the complex Higgs pole in the propagator is shown at infinite bare coupling in the O(4) model. The Higgs correlation length is calculated at each point.

A drastic change compared with Figure 4 of the large N result is immediately recognized. When the correlation length ξ is $O(1)$ the width of the Higgs particle is only about 20 percent of the heavy mass and we are still in the perturbative regime

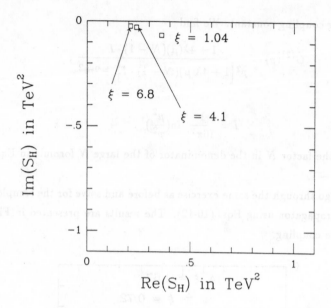

Figure 6. The position of the complex Higgs pole in the propagator is shown in the O(4) model at $\lambda(\Lambda) = 10$.

with a well-defined and reasonably sharp Higgs resonance pole. A similar picture at bare coupling $\lambda(\Lambda) = 10$ is depicted in Figure 6.

The resolution of the apparent contradiction is rather obvious. One will notice the factor $N + 8$ in the denominator of Eq. (10) which is the only difference in comparison with the large N formula of Eq. (4) (the deviation of the propagator formula in Eq. (11) from large N is relatively small). If we used the large N formula of Eq. (4) without checking its validity for $N = 4$, we would overestimate the renormalized coupling $\lambda(\mu)$, and therefore the Higgs mass bound, for a fixed correlation length ξ by a factor of 3! This would lead to the mistaken impression that the O(4) model, based on large N calculations, has a strongly interacting Higgs sector for heavy Higgs masses, close to 1 TeV.

Our results, based on a simple renormalization group analysis, are consistent with the direct application of perturbation theory to the lifetime calculation of the heavy Higgs particle in the electroweak model, or the O(4) approximation. Partial decay widths to various channels are shown in Figure 7.

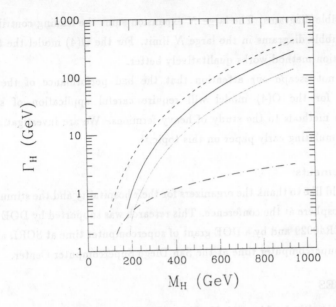

Figure 7. Partial decay widths of the $h \rightarrow t\bar{t}$ channel (dashed-dotted line), $h \rightarrow ZZ$ channel (dotted line), $h \rightarrow WW$ channel (solid line) are shown in the electroweak model as the function of Higgs mass m_H. The dashed line corresponds to the decay width of the Higgs particle into Goldstone particles in the $O(4)$ model.

5. CONCLUSION

Based on the above calculations and simulation results the following conclusions can be reached:

1. We have found no evidence that the $O(4)$ model has a strongly interacting region with a heavy Higgs particle for correlation lengths $\xi \geq 2$. For smaller correlation lengths the coupling constant (m_H/v ratio) is levelling off, without any signal for strong Higgs dynamics.

2. The large N expansion is not a good approximation for the $N = 4$ case in the $O(4)$ model which is physically relevant in the study of the electroweak model. The large difference between $N + 8$ and N for $N = 4$ is mainly responsible for the drastic change in the picture which has been recently advocated.[4]

3. Renormalization group improved perturbation theory with the running coupling constant is equivalent to summing all the leading logarithms from

all possible diagrams. The large N expansion sums the leading contribution from bubble diagrams in the large N limit. For the O(4) model the former summation method works qualitatively better.

It did not escape our attention that the bad performance of the large N expansion for the O(4) model will require careful application of similar approximation methods to the study of heavy fermions. We are investigating the results of a stimulating early paper on this topic.[5]

Acknowledgements

We would like to thank the organizers for their hospitality and the stimulating workshop atmosphere at the conference. This research was supported by DOE grant DE-AT03-81-ER40029 and by a DOE grant of supercomputer time at SCRI, and an NSF grant of supercomputer time at the San Diego Supercomputer Center.

REFERENCES

1. J. Kuti with L. Lin and Y. Shen, Non-Perturbative Lattice Study of the Higgs Sector in the Standard Model, this Proceedings, U.C. San Diego preprint UCSD/PTH 88-05, June 1988 (this paper contains a more comprehensive reference list on the topic of heavy Higgs mass bound).

2. Y. Shen with J. Kuti, L. Lin, and S. Meyer, Renormalized Perturbation Theory and New Simulation Results in the Φ^4 and O(4) Models, this Proceedings, U.C. San Diego preprint UCSD/PTH 88-06, June 1988.

3. G. Bhanot and K. Bitar, Regularization Dependence of the Lattice Higgs Mass Bound, this Proceedings, FSU-SCRI-88-50 preprint, June 1988.

4. M. B. Einhorn and D. N. Williams, On the Lattice Approach to the Maximum Higgs Boson Mass, this Proceedings, University of Michigan, Ann Arbor preprint, May 1988.

5. M. B. Einhorn and G. Goldberg, Phys. Rev. Lett., 57, (1986) 2115.

EXTRACTING PHYSICS FROM THE LATTICE HIGGS MODEL

Herbert Neuberger

Department of Physics and Astronomy
Rutgers University
Piscataway, NJ 08855

ABSTRACT

The relevance and usefulness of lattice ϕ^4 for particle physics is discussed from older and newer points of view. The talk will start with a review of the main ideas and suggestions in my work in the past with Dashen and will proceed to present newer developments both on the conceptual and the practical level.

In a paper published five years ago Dashen[1] and myself were the first to raise the question of how the triviality of ϕ^4 would affect the predictive power of the minimal standard model. We pointed out that:

(a) For small λ_R the model is a perfectly adequate effective theory. This is an important point to remember: it means that for weak coupling both the issue of triviality and, with it, the whole field of "lattice Higgs physics" may be irrelevant to particle physics.

(b) Experimentally, as far as we know, both the gauge and the Yukawa couplings are relatively small. This means that it would make no practical sense to try to include them in lattice calculations. [1] Thus it is only the ϕ^4 coupling which could be strong and for which nonperturbative techniques might be needed. In practice this means that we should only latticize the scalar sector and all of the remaining physics of the minimal standard model can be incorporated by ordinary perturbation theory. The sort of perturbation theory we are after is the following: express the quantity of interest as a series in

---

[1] The single plausible exception is the existence of a very heavy family - this possibility will be henceforth ignored.

the gauge and Yukawa couplings with coefficients given in terms of pure ϕ^4 expectation values. The simplest example is the "mass" of the vector boson $W^{2]}$ (evaluated from the zero momentum W-propagator):

$$M_W^2 = g^2 f^2/4 + O(g^4) \tag{1}$$

The pure $\lambda\phi^4$ quantity, f, is in the language of the Gell-Mann Levy linear sigma model the pion decay constant. It is free of renormalization arbitrariness because it is defined in terms of the currents:

$$\left\langle 0|J_\mu^\alpha(x)|\pi^\beta(k)\right\rangle = i \; f \; k_\mu \delta^{\alpha\beta} e^{-ikx} \tag{2}$$

for $k^2 = 0$, in Minkowski space, with properly normalized one-pion states on the right. The currents J are normalized by the current algebra ($\int d^3x J_0^\alpha(x) = Q^\alpha$ and the charges Q close a non-Abelian Lie algebra). By standard current algebra techniques one can easily show that:

$$\begin{aligned} \langle 0|\sigma|0\rangle &= z \; f \\ \left\langle 0|\pi^\alpha(x)|\pi^\beta(k)\right\rangle &= z\delta^{\alpha\beta} e^{-ikx} \end{aligned} \tag{3}$$

The relative normalization of the π and σ is fixed because they are members of the same multiplet and therefore the relations (3) can be taken over for the *bare* fields. Another example is the Higgs mass: it is given by the mass of the $I = J = 0$ state in the pure ϕ^4 model with subleading gauge and Yukawa corrections.

(c) The effective theory which works so well for small λ_R ceases to have a meaning when the ϕ^4 coupling becomes too strong (essentially when the unitarity bound gets violated). Triviality means that unitarity cannot be satisfied by just including higher order terms in the λ-expansion; a larger set of parameters have to be added into the theory in order to have the potential of producing an acceptable S-matrix. Thus the regime of applicability of the standard model is intrinsically bounded by the requirement of mathematical consistency, essentially in the same way an asymptotic series for a given function would become useless for too large values of the expansion parameter. In our case triviality means that, although we have an asymptotic series (perturbation theory), there does not exist a physically acceptable function—of the expansion variable only—whose series that is. More precisely: imagine that the "true" physics tells us that some observable O is given by a function $O(\lambda, g, \ldots, G_{new})$; we succeed, via ordinary renormalization theory, in defining an expansion in λ_R for which, to any finite order, there is no dependence on G_{new}. The asymptotic series so generated does not necessarily represent a reasonable function of λ_R only.

(d) To better understand the way triviality comes into the picture, it was suggested in my work with Dashen to look at: *1)* Large N generalizations of the $O(4)$ model. We cautioned that because of the well known perturbative $1/(N+8)$ terms the results so obtained would be of illustrative use only. Many people have since used $1/N$ to elucidate various points of view. I still have not given up the hope that somebody will come up with an expansion scheme which will be sufficiently close to the ordinary large N expansion to be tractable but which would include the $1/(N+8)$ terms in a natural way. There are quantum mechanical analogues to such things.[3] *2)* Approximate renormalization group recursion relations developed by Wilson. These methods are again reliable only on the illustrative level. P. Hasenfratz and various collaborators[4] have done extensive work in this direction. It is probably fair to say that no surprises have come out and, our original suggestion worked out to what one would have expected.

(e) Our main thrust in the paper was to try to set up a numerical study using Monte Carlo methods on the lattice for finding the point, in a given regularization, where the Higgs mass cannot be increased any further without becoming larger then the uv cutoff itself.

This approach was not without problems and some of them were pointed out in the original paper. There are two classes of problems one has: one class has to deal with issues of principle and the other with technicalities necessary for a useful practical implementation.

Principle: Obviously the bound will be scheme dependent (this was pointed out in the original paper[1] and in my talk at Seillac;[5] the first to investigate this in practice are Bhanot and Bitar in a very recent paper).[6] It is also clear that a "higher level" bound of bounds does not exist. So we have on our hands a bound which can be arbitrarily large; not something that would gain the respect of even the most generous mathematical physicist. Clearly this just reflects the fact that we deal with the minimal model where triviality means that we are dealing only with an asymptotic series without the backing of the function of which this series is an expansion. If we have a cutoff scheme, it is only the asymptotic series part of the result that is scheme-independent.

Nevertheless we do have a feeling that "reasonable" cutoffs will give us bounds of similar magnitude. To make further progress we need to focus on questions of more physical significance and establish (a) A better contact with the real world; i.e., what measurable process are we considering? It does not need to be something which would be directly related to the Higgs mass. Triviality holds for all aspects of the theory.[2] (b) And what do we mean by "reasonable."

In my opinion what has to be found out is whether a window of energies[5] exists between the Higgs mass (say) and the threshold for new physics (the composite scale for instance) which is so narrow that perturbative techniques are numerically impotent but strongly interacting ϕ^4 is still relevant as a description of the physics. One way to answer this question is to parametrize the higher physics by some dimension-6 terms in the ϕ^4 action with couplings measured in the scale of the new physics. Next one has to go to a regime where the Higgs is so heavy that ordinary perturbation theory is not applicable. If one can find there a region where the variation in the physical observables induced by changing the higher dimensional couplings in intervals of order unity is only a few percent of the total magnitude, the answer to the question is yes. This does not mean that that is what happens in the real world; it only means that such a possibility exists theoretically. It may turn out that the answer is negative and that such a possibility is ruled out by the ϕ^4 theory itself and that would basically mean that further lattice research of the scalar sector of the minimal model is irrelevant for particle physics. The main point is that the question can be asked and answered within ϕ^4 itself. Moreover, if one is willing to give up the traditional hypercubic lattice in favor of the F_4 lattice, Lorentz invariance is protected by the renormalization group up to and including dimension-6 operators. This means that there does exist a latticized version of ϕ^4

---

[2] Point (1) above can be restated in a different manner: what do we have to tell the world on the basis of Monte Carlo techniques that (a) they didn't know otherwise and (b) would be useful to know.

including dimension-6 operators which is potentially as close to the real world as hypercubic lattice field theory is in the strictly renormalizable case.[7]

A different way of thinking about the window is to seek a region where, although asymptotic scaling as given by the two-loop β-function no longer holds, one still has approximate, nonperturbative, scaling. This should remind us of lattice gauge theories. The work of the San Diego group[8] may be taken to imply that such a window has already been excluded. However, since their analysis has not allowed for a search for nonasymptotic scaling, I feel that the question about the existence of the window has not been answered yet even on the hypercubic lattice.

Maybe I should give some illustration of what kind of things except "Higgs masses" I have in mind: (a) The ρ-parameter; (in some definitions this parameter deals with the zero momentum propagators—we should remember that it is only at this kinematical point that direct contact can be made between the real, (Minkowski) world and the Euclidean space). (b) Try to generalize the N/D method for extrapolating from the Euclidean regime into the relevant kinematical regions. Use Monte Carlo to evaluate some of the coefficients which have to be input into the extrapolation. (c) Look at other states with different I, J quantum numbers.

Practice: The ϕ^4 system is in the broken phase and the presence of Goldstone particles causes severe finite size effects. This issue was mentioned in the original paper but the proposal there was not very good on the practical level, and has, in practice, never been satisfactorily implemented. Recently, significant progress was made on this technical issue with the help of soft-pions finite-size theorems.[9]

I should be careful to explain what is meant by this term. If a nontrivial continuum limit existed we would go to the limit $f \to 0$ and $L \to \infty$ with $fL = l$ kept fixed. We would then discuss the dependence on l of various observables. To leading order in the typical momenta (also measured in units of f) we would expect to be able to evaluate the l-dependence without any additional information (this is a consequence of soft-pion physics being fixed by symmetry considerations alone). Here we need something slightly (but clearly) different: we want the leading L dependence of some observables evaluated still on the lattice (in the bare theory) at fixed couplings. In particular we have to measure f which is a function of the couplings defined at $L = \infty$; this is the quantity which will set the scale for any

202

other dimensional quantity we may consider. The main formula is:

$$\left\langle \vec{\phi}\vec{\phi} \right\rangle_L = z^2 \left[f^2 - \frac{(N-1)\alpha}{L^2} \right] + O\left(\frac{\log^\gamma L}{L^4} \right) \qquad \vec{\phi} = L^{-4} \sum_x \vec{\Phi}(x) \qquad (4)$$

where the boundary conditions are periodic. α is a universal number, dependent only on the shape of the box and it can be calculated for boxes associated with the symmetry of hypercubic and F_4 lattices. While there is little doubt that the formula is correct, a complete and detailed proof has not been given yet.

Urs Heller and I[9] have perfected a way to extract f which makes use of the formula and also provides a check for it. The essential new ingredient is to devise a way to deal with the corrections subleading to the main formula. For large volumes the quantity $\vec{\phi}$ will be distributed with a probability density which is very sharply peaked.[3] Therefore, the moments of $(\vec{\phi})^2$ will be calculable in a very good saddle point expansion with leading corrections of the order $\log^\gamma L/L^4$. Therefore the theorem tells us that the center of the probability distribution (the most probable value, a_L^2) is the one which gets shifted by an $O(L^{-2})$ term. The piece of the measured $(\vec{\phi})^2$ which comes from the saddle point expansion can be extracted by measuring higher moments, i.e., from the data itself. There are also $\log^\gamma L/L^4$ corrections in a_L^2 which are well approximated by using an "effective pion mass." These terms are only important for really small volumes. A graph of a_L^2 as a function of the theoretically predicted finite size correction shows a striking linear dependence. From the slope we get z^2 and from the intercept we obtain f^2.

The above gives us a good way for measuring f. This by itself is nontrivial: as a matter of fact there is no measurement of f that I am aware of, in spite of the rather significant amount of activity on the subject. Most workers have been using an estimate of $\langle 0|\sigma_R|0\rangle$ instead. This is neither easier to measure nor the correct quantity in eq. (1). The scale in the broken sector of ϕ^4 is set by f. The claim that one is assured that $\langle 0|\sigma_R|0\rangle^2 \approx f^2$ is not well founded as Urs Heller and I[12] have

---

[3] If the logarithm of the probability density were really extensive it would be proportional to the volume. However, due to the long range of the interactions, the volume factor is softened by logarithms. This is just saying that the sigma propagator has a two-pion cut.

shown by looking at a three-dimensional example.[4]

The other issue is related to what else we measure. Here I shall restrict myself to discuss only the Higgs mass although I strongly suspect that it is not really a terribly good idea to concentrate on this quantity. We are after all interested only in the case where the Higgs particle has a mass of about 1 TeV and then its width would be about .5 TeV.

There are three different definitions and consequently different methods for getting the Higgs mass: (a) Look for exponential decay [(really cosh(...)+Const.) in a calculation of the "time-correlations"] of spatially averaged (zero space momentum) "sigma" fields.[11] (b) Take a "sigma" two point function as a function of four-momentum (Euclidean) and fit it to a free propagator with an undetermined mass and wave function renormalization constant.[8] (c) Extract the coefficient of the second variation from the logarithm of the probability distribution.[8] (This logarithm would, when $L \to \infty$ have a convex hull giving the effective potential times a volume factor; however, we want the coefficient at the miminum and the latter is infrared singular).

To understand what is meant in (a) and (b) above we need to define the σ-field. This is not free of ambiguity in a finite volume because the symmetry is not broken there. The problem is somewhat similar to the issues answered by "complementarity:" we must find an invariant operator (under the global O(N)) which has the property that, in perturbation theory, it looks like the σ-field (and the latter expansion is meaningful in the infinite volume limit):

$$\sigma(x) = \vec{\phi}\vec{\Phi}(x)/|\vec{\phi}|$$
$$\vec{\pi} = \vec{\Phi} - \vec{\phi}\sigma(x)/|\vec{\phi}| \tag{5}$$

It is clear now that (c) above simply proposes to extract the mass from the zero momentum piece of the σ two point function. We therefore shall think only about (a) and (b). It is also quite obvious that, if (b) works (i.e., the fit to a free propagator is good) then (a) and (b) should give the same answer. However (a) is safer because, in principle, it will give an estimate for the energy of the lowest excited state with quantum numbers $I = J = 0$ and this is true regardless of anything else.

---

[4] That the z factor in eq. (3) is not unity can be seen from the explicit calculations of B. Lee in ref [8], specifically equations (31), (33) and (34) there.

But it is clear that if the volume is large enough, the lowest state in the Higgs channel is the two-pion state. As long as the volume is finite, the two-pion cut is represented by a sequence of poles at the appropriate discrete values of the energy determined by the momenta of the pions. Now let's do some simple arithmetic: if the inverse Higgs mass is 1–2 lattice spacings (close to the bound), the Higgs particle will be lighter than the lowest two-pion state[5] when $\frac{4\pi}{L} > 1 - 2$ or $L < (2 - 4)\pi$. Therefore, for most practical volumes and interesting masses, the two-pion state of lowest relative momentum will strongly mix with the Higgs particle and whatever we measure may have little to do with the position of the complex pole in momentum-square of the σ two-point correlation function at infinite volumes. I strongly suspect that what I described means that there are very sizable finite size systematic errors in all the values quoted for the Higgs mass in the Monte-Carlo literature. I see no concrete solution to this problem. A (somewhat wild) speculation is that one should try to extract the multipion part theoretically by some sort of new finite-size soft-pion theorem, try to adapt Lüscher's finite volume analysis to resonances and then apply it to the Higgs particle.

But we shouldn't complain about the Higgs trouble. It really is not less of a problem in real physics; a broad resonance has no distinct signature. Since we are ultimately interested in making an experimentally relevant statement, it makes little sense to bother with the Higgs mass so much. Before we get to the real thing we would still like some measure of the strength of the interaction which has some sort of simple theoretical meaning. The simplest candidate is obviously the coupling constant; the strength of the two-pion cut in the sigma propagator should provide a possible definition for it. Essentially one has to see the $\log(L)$ piece in the zero momentum sigma propagator. This is, however, difficult to see mainly for the same kinematical reasons which made the Higgs mass measurement unreliable. But, in my opinion, it is better to think that you are trying to get a signal of the cut and an estimate for its strength rather than an estimate for the location of a pole far off in the complex plane.

If kinematics is more on our side it may be easier to see the cut rather than to chase after the Higgs mass. The cut is clearly hidden in the large volume behavior

---

[5] There are no pions at strictly zero momentum with our definition of the pion field.

of the $|\vec{\phi}|$ probability distribution in the vicinity of its saddle:

$$p(x)dx = N\exp[-B_L(x-1)^2 + O((x-1)^3)]\,x\,dx \quad x = \phi^2/a_L^2. \tag{6}$$

If the sigma were stable B_L would go like the volume. The two-pion cut makes it go like the volume devided by $\log(L)$ and the log is difficult to detect numerically. To improve our chances we go to a lower dimension where infrared problems must be enhanced. In three dimensions the two-pion cut which was logarithmic in four dimensions becomes a square root cut.[6] This means that B_L will now go as L^2 (symmetric box assumed) instead of L^3. In collaboration with U. Heller this behavior has indeed been observed.[12]

We would like to be able to do something similar in four dimensions; to see the logarithm will be difficult. We could try to look for a "lattice two-pion cut" given by $L^{-4}\sum_{r\neq 0} 1/\{4\sum_\mu \sin^2[\pi r_\mu/L] + m_{\text{eff}}^2\}^2$ (with m_{eff}^2 going like $1/V$). I think this is not hopeless but still not quite feasible and I would like to explain why. B_L is clearly related to $Var((\vec{\phi})^2)$. Let's calculate the latter quantity in the $1/N$ expansion. We find:

$$Var((\vec{\phi})^2) = \frac{2N}{L^4}\frac{X}{1+(m_L^2 L^4)^2 X/L^4} \tag{7a}$$

where

$$X = \frac{1}{g_0^2} + L^{-4}\sum_{r\neq 0} 1/[4\sum_\mu \sin^2(\pi r_\mu/L) + m_L^2]^2 \tag{7b}$$

and, for large L we can take $m_L^2 = \dfrac{N}{L^4\langle\sigma\rangle^2}$. Clearly the quantity of interest is X. We would like to write X as $X = \dfrac{1}{g_{\text{phys}}(f)} + \dfrac{1}{8\pi^2}\log(fL)$ where g_{phys} has been defined at a scale f. But we see that part of the loop giving the two-pion cut, namely the high momentum part, has to be incorporated into the first term to renormalize the coupling. There is no doubt that this renormalization is sizable; very rough estimates for g_{phys} show that it is significantly smaller than g_0 so that a large piece of the lattice loop goes into the coupling and little is left for representing the two-pion cut. Some rough calculations I carried out have led me to estimates for the volume sizes needed in order to see the cut which are too depressing to be appropriate for

---

[6] Certainly the three dimensional theory is interacting in its continuum limit but this has little to do with what we can see or not on the lattice.

206

mentioning here. In short, the problem is that we need a momentum space large enough to allow some sort of separation between hard and soft momenta.

In closing I should remind you that a self-consistent theory which may have a Higgs-system resembling low energy sector does exist and is well known: it is QCD. The ϵ is not an established resonance; however, if it does exist it has a mass of about 700 MeV. The pion decay constant is 93 MeV. Neglecting all but u and d quarks we could say that while it would be stretching it to claim that a meaningful regime for an effective ϕ^4 exists in QCD, nevertheless, the Higgs bound cannot be too grossly overestimated. This gives us, in weak interaction units, a Higgs mass upper bound of 2 TeV. In view of this I find it hard to believe that the real bound is as low as 550 GeV or even 650 GeV.

Acknowledgements

I am grateful to G. Bhanot and U. Heller for discussions. This research was supported in part by the DOE under the OJI program grant number DE-FG05-86-ER40265.

REFERENCES

1. R. Dashen, H. Neuberger, Phys. Rev. Lett.50, (1983) 1897.

2. See for example E. Farhi, L. Susskind, Phys. Rep.74, (1981) 277.

3. L. D. Mlodinow, N. Papanicolaou, Ann. Phys.128, (1980) 314 and 131 (1981) 1.

4. P. Hasenfratz, J. Nager, Z. f. Phys. C (1988) 477 and references therein.

5. H. Neuberger, Nucl. Phys. B(Proc. Suppl)4 (1988) 501.

6. G. Bhanot, K. Bitar, FSU-SCRI-88.

7. H. Neuberger, Phys. Lett. 199B, (1987) 536.

8. J. Kuti et al. UCSD/PHH 87-19

9. H. Neuberger, Phys. Rev. Lett. 60 (1988) 889; Nucl. Phys. B, to appear; U. M. Heller, H. Neuberger, Phys. Lett. B, to appear.

10. B. W. Lee, Nucl. Phys. B9 (1969) 649.

11. A. Hasenfratz et al., Phys. Lett. 199B, (1987) 531.

12. U. M. Heller, H. Neuberger, work in preparation.

EXCITED STATES OF GAUGE MODELS

Janos Polonyi†**

Center for Theoretical Physics
Laboratory for Nuclear Science
and Department of Physics
Massachusetts Institute of Technology
Cambridge, Massachusetts 02139 U.S.A.

ABSTRACT

It is shown that matrix elements computed between the physical states in gauge theories have one of the following features: Particles carrying center charges are either confined or propagating but their charge is screened completely by some unusual states of the gauge field. Thus the center charge is permanently confined. The deconfined screened matrix elements support particles with unusual spin-statistics.

1. INTRODUCTION

This work deals with some nonperturbative phenomena of $SU(N)$ Yang-Mills field coupled to matter field which transforms nontrivially under the center of global gauge transformations. In case of fermions the coupling is assumed to be vector-like without gauge anomalies. We shall study the matrix elements of the matter propagator in an arbitrary gauge. It will be shown that either (i) these matrix elements vanish or (ii) the center charge of the matter particles is screened by some states of the gauge field. We interpret the vanishing of the matter propagator as the impossibility of creating an isolated particle. Thus the confinement of the center charge is permanent. The thermal averages computed in a high temperature Yang-Mills system exhibit the screened properties. Since these averages are the mean values of expectation values over an energy shell $E(T)$ when the microcanonical ensemble is used, the screened properties are typical of the highly excited states.

---

† Alfred P. Sloan Research Fellow
** On leave of absence from CRIP, Budapest, Hungary

The states which screen the matter particles must be unusual since the gauge field itself is invariant under the center of the global gauge group. The construction of charged states by neutral operators is achieved by exploiting the multiconnected topology of the space of global gauge transformations. Magnetic monopoles are introduced as typical gauge-hedgehog configurations. It is shown that such particles acquire electric charges and appear as fermions in the screened phase. Due to the permanent confining forces between the matter particles and the magnetic monopoles new bound states are formed. The matter particles appear as composite objects with changed spin-statistics.

We discuss the case of the quenched quark propagator first. The bosonic case and the contributions of the dynamical loops will be considered at the end only. Our discussion is the sytematic reconstruction of the usual confinement-deconfinement phase transition[1] in temporal gauge operator formalism. Our argument holds for any finite value of the ultraviolet cutoff in regularizations which preserve local gauge invariance.

2. THE PROJECTION OPERATOR OF YANG-MILLS THEORIES

The $SU(N)$ Yang-Mills Hamilton operator is of the form

$$H = -\operatorname{tr} \int \left(g^2 \vec{E}^2(\vec{x}, t) + \frac{1}{g^2} \vec{B}^2(\vec{x}, t) \right) d^3 \vec{x} , \tag{1}$$

in temporal gauge $A_0 = 0$ where $\vec{A}(\vec{x}) = \vec{A}^a(\vec{x}) \frac{\lambda^a}{2i}$, $\vec{E}(\vec{x}) = \vec{E}^a(\vec{x}) \frac{\lambda^a}{2i}$, $E_i^a(\vec{x}) = -i \frac{\delta}{\delta A_i^a(\vec{x})}$ and $B_i = \partial_j A_k - \partial_k A_j + [A_j, A_k]$. The canonical commutators give

$$\left[\left(\vec{D} \vec{E}(\vec{x}) \right)^a, \left(\vec{D} \vec{E}(\vec{y}) \right)^b \right] = 2i f^{abc} \left(\vec{D} \vec{E}(\vec{x}) \right)^c \delta(\vec{x} - \vec{y}) , \tag{2}$$

and

$$[H, \vec{D} \vec{E}(\vec{x})] = 0 . \tag{3}$$

One has to introduce counterterms in (1). All what we need to know is that they preserve gauge invariance and (3) remains valid. Gauge transformations act as $\vec{A}^g = g(\vec{\partial} + A)g^\dagger$ and physical states carrying center N-ality charge n are obtained by means of the projection operator $\mathcal{P}_n = \mathcal{P} P_n$

$$\mathcal{P} = \mathcal{N}^{-1} \int \mathcal{D}_{\mathcal{G}/Z_N} [t A_0(\vec{x})] e^{it \int \operatorname{tr} \left(A_0 \vec{D} \vec{E} \right) d^3 \vec{x}} , \tag{4}$$

and

$$P_n = \frac{1}{N} \sum_{\ell=1}^{N} e^{-i\frac{2\pi}{N}n\ell} z^\ell \quad .(5)$$

The integral in (4) involves the invariant measure $D[tA_0(\vec{x})]$ and extends over the direct product $\mathcal{G}$ of the local gauge groups divided by the center Z_N of the global gauge transformations. The function $A_0(\vec{x})$ should obey some smoothness conditions as well if continuum regularization is used. The z of (5) is the global gauge transformation by the center element $e^{\frac{2i\pi}{N}}$. The projection operator can be written as

$$\mathcal{P}_n = \mathcal{N}^{-1} \int \mathcal{D}_\mathcal{G} \left[tA_0(\vec{x})\right] e^{i \int \, \mathrm{tr} \left(tA_0\vec{D}\vec{E}\right)d^3\vec{x} - \frac{in2\pi}{N}\ell[tA_0]} \, , \qquad (6)$$

$$\ell\,[\alpha] = M \left(\mathrm{Im} \log \, \mathrm{tr} \frac{1}{V} \int d^3x \, e^{\alpha(\vec{x})} \right) \, , \qquad (7)$$

where the function $M(y)$ gives the closest real number to y which has the form $2\pi k/N, k$ integer. The states prepared by $\mathcal{P}_n$ give

$$z\mathcal{P}_n|\rangle = e^{\frac{2i\pi n}{N}} \mathcal{P}_n|\rangle \quad .(8)$$

States with nonvanishing N-ality can be made by the help of the gauge field. The point is that the functional space $\left\{\vec{A}(\vec{x})\right\}$ is multiconnected as far as global gauge transformations are concerned. In fact, the gauge field changes as $\vec{A} \to g\vec{A}g^\dagger$ under global gauge transformations which gives the group space $SU(N)/Z_N$. The operators $\mathcal{P}_n, 0 < n < N$ select the states whose wave functionals are multivalued when global gauge transformations are performed. These unusual states of the gauge field possess electric charge which may screen the matter fields.

The projection operator which prepares the physical states with an external charge in the multiplet r at location $\vec{x}_0$ is given as

$$\mathcal{P}[r,\vec{x}_0] = \mathcal{N}^{-1} \int \mathcal{D}_\mathcal{G} \left[tA_0(\vec{x})\right] e^{i \int \, \mathrm{tr} \left(tA_0\vec{D}\vec{E}\right)d^3\vec{x}} \, \mathrm{tr} \, e^{-tA_0^a(\vec{x}_0)\frac{\lambda_r^a}{2i}} \, , \qquad (9)$$

where the generators λ_r^a act in the representation r of the gauge group.

One performs integration by part in the exponent of (6) and (9) in deriving the path integral expressions. It is important to note that the surface term is to be neglected in this step. To see this, consider $U(1)$ gauge theory in lattice regularization. The effect of the gauge transformation characterized by the function $\alpha(\vec{n})$ on the state functionals is

$$\psi\left[\vec{A}(\vec{n})\right] \to \psi\left[\vec{A}(\vec{n}) + \vec{\nabla}\alpha(\vec{n})\right] , \tag{10}$$

where the gradient contains the link contributions of the lattice. This transformation is generated either by the operator $e^{i\sum \vec{E}\vec{\nabla}\alpha}$ or $-i\sum \alpha\vec{\nabla}\vec{E}$ as long as links of the lattice appear only in the summation.

The situation is slightly more involved for non-Abelian gauge groups due to the connection term in the covariant derivative. Though the commmutation relations guarantee the correct behavior for the operator $e^{\int \alpha\vec{D}\vec{E}}$ this is not obvious for $e^{-\int \vec{E}\vec{D}\alpha}$. The path integral in lattice regularization directly uses the matrix elements of the operator which generates the nonabelian analogue of (10). The equivalence of the two types of transformations can be easily seen in the continuum case, first for gauge transformations which can be obtained from the identity by repeated application of infinitesimal steps. We consider the one-parameter family of gauge transformations $g_\tau(\vec{n}) = e^{\tau\alpha(\vec{n})}, 0 < \tau < 1$ and the differential equation

$$\frac{d}{d\tau}|\tau\rangle = \int \vec{E}\vec{D}\alpha|\tau\rangle , \tag{11}$$

which is satisfied by the states $|\tau\rangle = |\vec{A}^{g_\tau}\rangle$ and $e^{\int \vec{E}\vec{D}\alpha}|\vec{A}\rangle$. Since these states obey the same initial condition at $\tau = 0$ they agree at $\tau = 1$. The extension of this result for gauge transformations with a nonvanishing Pontryagin index is obtained by repeating the construction (4) for the group $\mathcal{G}/Z$ where the factor Z is the homotopy group corresponding to gauge transformations with nonzero winding[2]. The projection operator P_Θ is obtained as in (5) except replacing $\frac{2\pi}{N}n$ by Θ and z by a gauge transformation with unit winding.

The propagator in the zero external charge sector is given by the matrix element

$$\left\langle \vec{A}^{(f)} \left| \mathcal{P}_0\, e^{-itH} \right| \vec{A}^{(i)} \right\rangle , \tag{12}$$

which is written as

$$\mathcal{N}^{-1} \int \mathcal{D}_{\mathcal{G}}\left[tA_0(\vec{x})\right] \left\langle \vec{A}^{(f)} \left| e^{-itH_{A_0}} \right| \vec{A}^{(i)} \right\rangle , \tag{13}$$

where

$$H_{A_0} = -\operatorname{tr} \int \left\{ g^2\vec{E}^2(\vec{x}) + \frac{1}{g^2}\vec{B}^2(\vec{x}) + A_0(\vec{x})\vec{D}\vec{E}(\vec{x}) \right\} d^3\vec{x} . \tag{14}$$

This formalism allows us to gain some insight into the dynamics of the matrix element of $e^{-itH_{A_0}}$ between unconstrained states. One postpones imposing Gauss's law in this manner.

3. THE CONFINED PHASE

Consider the quenched quark propagator

$$\mathcal{A} = \left\langle \vec{A}^{(f)} | T\psi(y)\bar{\psi}(x) | \vec{A}^{(i)} \right\rangle_{phys} = \left\langle \vec{A}^{(f)} | \mathcal{P}_0 T K^{-1}[x,y; \vec{A}, A_0] e^{-itH} | \vec{A}^{(i)} \right\rangle , \quad (15)$$

and insert[3] z

$$\mathcal{A} = \left\langle \vec{A}^{(f)} | z \mathcal{P}_0 T K^{-1}[x,y; \vec{A}, A_0] e^{-itH} | \vec{A}^{(i)} \right\rangle$$

$$= \mathcal{N}^{-1} \int \mathcal{D}_{\mathcal{G}}\left[tA_0\right] \left\langle \vec{A}^{(f)} | z T K^{-1}[x,y; \vec{A}, A_0] T e^{-itH_{A_0}} | \vec{A}^{(i)} \right\rangle$$

$$\quad (16)$$

$$= \mathcal{N}^{-1} \int \mathcal{D}_{\mathcal{G}}\left[tA_0^z\right] \left\langle \vec{A}^{(f)} | T e^{-i\frac{2\pi}{N}} K^{-1}[x,y; \vec{A}, A_0^z]) T e^{-itH_{A_0^z}} | \vec{A}^{(i)} \right\rangle$$

$$= e^{-i\frac{2\pi}{N}} \mathcal{A} .$$

The first equation holds when z acts to the left since $\vec{A}^z = \vec{A}$. The last equation is obtained by applying z to the right when the the change of variable

$$e^{tA_0^z(\vec{x})} = z e^{tA_0(\vec{x})} \quad (17)$$

is allowed. The phase factor $e^{-i\frac{2\pi}{N}}$ is to compensate the change $K^{-1}[x,y; \vec{A}, A_0^z] = e^{i\frac{2\pi}{N}} K^{-1}[x,y; \vec{A}, A_0]$. This chain of equations is satisfied only if $\mathcal{A} = 0$. The propagator vanishes as the result of the completely destructive interference between the homotopically different trajectories in the space of global gauge transformations[3]. This is the case (i) announced in the Introduction.

The presented argument is valid for periodic boundary conditions for $A_0(\vec{x})$ in the three-space which is used in lattice computations. Semiclassical arguments can be applied to restrict the domain of integration in the projection operator for the function space $A_0(\vec{x}) \to A_0$ as $|\vec{x}| \to$. The boundary conditions where A_0 is held at a fixed value exclude the change of variables (17). We introduce a local gauge transformation in this case which gives a center element z within a sphere with radius R and approaches e^{tA_0} as $|\vec{x}| \to$. The matrix element is called confining when our local gauge transformation becomes a symmetry and (17) is allowed as

$R \to$, *i.e.* when the symmetry breaking boundary conditions are irrelevant in the thermodynamical limit. It remains to be proven that the vacuum expectation values belong to this class.

4. DECONFINED-SCREENED PHASE

The change of variable (17) is not allowed when the infinite dimensional integration of the projection operator does not include the summation over the center of global gauge transformations. This phenomenon is similar to the ordinary dynamical symmetry breakdown. Note that the symmetry in question is precisely the one which is found to be broken dynamically in the path integral studies of the deconfinement phase transitions. The order parameter for the symmetry transformation is $e^{-\beta F_q}$ where $\beta = 1/T$ and F_q is the energy of a static test quark. The screened phase is realized by the projection operator

$$\mathcal{P}_{dec} = z^0 \mathcal{P} = \sum_{k=1}^{N} P_k \mathcal{P} \ . (18)$$

The first expression is useful in the field diagonal basis and reflects the fact that the summation over A_0 is restricted into one Z_N sector only. It is advantageous to use the charge diagonal representation where the states are classified according their transformation properties under the center of the global gauge group. The second equation is used in this representation where it expresses the fact that the absence of the summation of the projection operator allows the presence of charged states.

The propagator $\mathcal{A}$ which serves as an order parameter for the center is nonvanishing in this case. The repetition of the chain of arguments of equation (16), when $\mathcal{P}_{dec}$ is used, shows that nonzero contributions for $\mathcal{A}$ may come from the action of the operator P_1 only which compensates the phase generated by the matter propagator. Thus quarks are bound permanently to some kind of gluon states in the deconfined phase. These gluon states screen the center charge of the quarks.

5. MAGNETIC MONOPOLES

Any nonzero gauge field configuration $\vec{A}(\vec{x})$ can be used to construct states carrying center charge. In fact, the linear combination

$$\int dg \ \psi(g) |g\vec{A}(\vec{x})g^\dagger\rangle \ , \tag{19}$$

where the integration is over the global gauge group, transforms nontrivially under the center when the wave function $\psi(g)$ is multivalued on $SU(N)/Z_N$.

If the space and the gauge directions are correlated in $\vec{A}(\vec{x})$, then the state has rather peculiar properties. Consider the prototype of such a gauge-hedgehog in $SU(2)$ Yang-Mills theory

$$A_i^a(\vec{x}) = G(|\vec{x}|)\epsilon_{iaj}x^j , \tag{20}$$

which is similar to the gauge field corresponding to the 'tHooft-Polyakov monopole except the static Higgs field is replaced by $A_0(\vec{x})$. These configurations can be defined in a manifestly gauge invariant way since A_0 transforms homogeneously under local gauge transformations[4]. The peculiar feature of the configuration (20) is

$$gA_i(\vec{x})g^\dagger = \mathcal{R}(g^\dagger)_{ij}A_j(\mathcal{R}^{-1}(g^\dagger)\vec{x}) , \tag{21}$$

where $\mathcal{R}(g)$ is the adjoint representation of $SU(2)$. This equation follows from

$$\epsilon_{ijk} = \epsilon_{lmn}\mathcal{R}_{il}\mathcal{R}_{jm}\mathcal{R}_{kn} , \tag{22}$$

and indicates that the effect of a global gauge transformation can be compensated by the inverse spatial rotation for such configurations. The double valuedness of the wave functional of a gauge-hedgehog with respect to global gauge transformations leads to the same properties when spatial rotations are considered. Thus gauge-hedgehogs which have a double-valued wave functional have half-integer spin and behave as fermions according to the spin-statistics theorem[5].

This mechanism generalises to $SU(N)$ gauge theories. Since the magnetic monopoles of $SU(N)$ gauge models correspond to some $SU(2)$ subspace of the gauge group the states carrying center charge will obey ordinary fermi statistics rather than fractional ones. The theory contains N versions of each monopole in the deconfined-screened phase. One of these, which is the result of the action of P_0, remains boson and neutral. The other $N - 1$ species become dyon and acquire fermi statistics in agreement with Ref. [6].

6. MATTER FIELDS

The gauge-hedgehog configurations cause fundamental changes in the dynamics of the deconfined matter particles. Consider a matter particle which is screened

214

by a gauge-hedgehog. This particle appears with unusual spin-statistics properties borrowed from the permanently bound hedgehog. The matter particles can be screened by the state of the gauge field whose wave functional is nonvanishing for ordinary gauge field configurations and gauge-hedgehogs. Due to the superselection rules for the spin, either the bosonic or the fermionic component can give contributions in a given physical process. Thus properly chosen screened matrix elements receive contributions from the fermionic hedgehog components only.

Finally we consider the case of dynamical fermions. After integrating over the Grassman variables we are left with an effective action containing closed Wilson loops and open Polyakov line-type contributions. The latter arises from the integration over nonzero baryon number sectors and describes the propagation of the "background" quarks. We shall restrict ourselves to the case of isolated quarks which assumes vanishing "background" center charge[7]. The effective action contains those combinations of the Polyakov line-type contributions which remain invariant under the insertion of z in (16). Thus our argument remains valid.

In the case of bosonic matter field the matrix element of the matter propagator can be obtained by integrating out the matter field in perturbation expansion. The application of real-space Feynman rules yields the expression where the integration over the matter field generates the insertion of the interaction vertices and internal particle loops along the world line of the "valence" particle. Similar to the fermionic case, these contributons remain invariant under the application of z in the zero central charge sector as long as the interaction vertices are charge preserving.

Acknowledgements

This work is supported in part by funds provided by the U. S. Department of Energy (D.O.E.) under contract #DE-AC02-76ER03069.

REFERENCES

1. L. McLerran and B. Svetitsky, *Phys. Lett.* **98B**, 196 (1980); J. Kuti, J. Polonyi and K. Szlachanyi, *Phys. Lett.* **98B**, 199 (1980).

2. J. Polonyi, "Gauge invariance and quark confinement," in the Proceedings of the International Symposium on Space-Time Symmetries, University of Maryland, College Park, Maryland, May, 1988, to be published in *Nucl. Phys.* **B**.

3. J. Polonyi, "Quantum Mechanics on Multiply Connected Spaces and Quark Confinement," MIT Preprint, CTP# 1507.

4. J. Polonyi, *Nucl. Phys.* **A461**(1978), 279c. (1985).

5. D. Finkelstein and J. Rubinstein, *J. Math. Phys.* **9**, 1762 (1967).

6. R. Jackiw and C. Rebbi, *Phys. Rev. Lett.* **36**(1976)1116, P. Hasenfratz and G. 'tHooft, *Phys. Rev. Lett.* **36**(1976)1119, A. S. Goldhaber, *Phys. Rev. Lett.* **36**(1976)1122.

7. A. Roberge and N. Weiss, *Nucl. Phys.* **B275**(1986)734.

RENORMALIZED PERTURBATION THEORY AND NEW SIMULATION RESULTS IN THE Φ^4 AND O(4) MODELS

Y. Shen*   with J. Kuti, L. Lin, and S. Meyer†

Department of Physics
University of California at San Diego
La Jolla, CA 92093, USA

ABSTRACT

The effective potential and Euclidean Green's functions of the four component scalar field theory with O(4) symmetric self-interaction are investigated in renormalized perturbation theory. The infrared singularities of Euclidean Green's functions, generated by massless Goldstone modes in the broken symmetry phase, are discussed and their finite volume regularization in Monte Carlo simulations is described. New Monte Carlo results for the one-component Φ^4 model and the O(4) model are presented and compared with our theoretical calculations which include a renormalization group analysis of the effective potential. We find from our analytic and numerical studies that a strongly interacting Higgs sector with a very heavy Higgs particle is an unlikely scenario in the electroweak model.

1. EFFECTIVE POTENTIAL AND LOOP EXPANSION

The importance of the O(4) model in the study of the electroweak model was discussed in a previous talk at this conference.[1] We have seen there that the O(4) limit of the SU(2) Higgs model with lattice regularization is given by the Euclidean action

$$S = \frac{1}{2}\sum_i \left[\sum_{\hat{\mu}}(\vec{\Phi}_{i+\hat{\mu}} - \vec{\Phi}_i)^2 - m^2\vec{\Phi}_i^2\right] + \lambda\sum_i \vec{\Phi}_i^4 \, , \tag{1}$$

where m^2 and λ are bare parameters and the field variables $\vec{\Phi}_i$ are O(4) vectors on

---

* speaker at the conference
† affiliation: Dept. of Phys., University of Kaiserslautern

lattice sites labelled by i. The unit vector $\hat{\mu}$ points along the four positive lattice directions and the lattice spacing a is set to unity in our calculations (in some equations a will be displayed for the clarity of the presentation).

1.1. One-loop results

The effective potential $U_\Omega(\phi)$ is defined by

$$\exp[-\Omega U_\Omega(\vec{\phi})] = \int D[\vec{\Phi}]\, \delta(\vec{\phi} - \frac{1}{\Omega}\sum_i \vec{\Phi}_i)\, e^{-S[\vec{\Phi}]}\,, \tag{2}$$

where Ω designates the finite lattice volume (it is equal to the number of lattice sites in our units) and the integration is over the field variables $\vec{\Phi}_i$. The O(4) symmetric effective potential $U_\Omega(\vec{\phi})$ is non-convex in the broken symmetry phase and depends on $\phi = \sqrt{\vec{\phi}^2}$ alone.

One can develop a systematic loop expansion for the effective potential in a finite lattice volume.[2] The one loop-result for the general O(N) case is given by

$$U_\Omega(\phi) = V(\phi) + \frac{1}{2\Omega}\sum_{\hat{k}\neq 0} ln\left[\hat{k}^2 + V''(\phi)\right] + \frac{N-1}{2\Omega}\sum_{\hat{k}\neq 0} ln\left[\hat{k}^2 + V'(\phi)/\phi\right] \\ + \delta V(\phi)\,. \tag{3}$$

In Eq. (3) $V(\phi)$ is the tree level potential,

$$V(\phi) = -\frac{1}{2}m^2\phi^2 + \lambda\phi^4\,, \tag{4}$$

where $V'(\phi)$ is the first derivative of $V(\phi)$, and $V''(\phi)$ is the second derivative of $V(\phi)$, with respect to ϕ. The massless lattice propagator $\hat{k}^2$ is defined by

$$\hat{k}^2 = \hat{k}_\mu \cdot \hat{k}_\mu, \quad \hat{k}_\mu = \frac{2}{a}\, sin(\frac{1}{2}ak_\mu)\,, \tag{5}$$

where the lattice momentum k_μ with $\mu = 1,2,3,4$ takes discrete values appropriate for helical boundary condition on a finite lattice in our Monte Carlo simulations. On the infinite lattice the continuous range of k_μ is between $-\pi/a$ and $+\pi/a$.

In Eq. (3) counterterms $\delta V(\phi)$ render $U_\Omega(\phi)$ finite in the $a \to 0$ limit. The precise choice of $\delta V(\phi)$ in Eq. (6) will make the finite parameters m and λ unambiguously defined in the tree level potential $V(\phi)$ of Eq. (4). Since $\delta V(\phi)$ depends on the arbitrary scale parameter μ, the renormalized tree level parameters

m and λ have to be tuned with μ to preserve renormalization invariance of physical observables. Although the μ dependence is not indicated explicitly, the parameters $m(\mu)$ and $\lambda(\mu)$ in Eq. (4) should not be confused with the bare parameters of Eq. (1).

Since the finite lattice volume has no effect on the short distance behavior of Euclidean lattice Green's functions in the $a \to 0$ limit, $\delta V(\phi)$ is chosen to be identical to the infinite volume counter terms,

$$
\begin{aligned}
\delta V(\phi) = &-\frac{1}{2}V''(\phi)\int_{-\pi/a}^{+\pi/a}\frac{d^4k}{(2\pi)^4}\frac{1}{\hat{k}^2} + \frac{1}{4}\Big(V''(\phi)\Big)^2\int_{-\pi/a}^{+\pi/a}\frac{d^4k}{(2\pi)^4}\frac{1}{(\hat{k}^2+\mu^2)^2} \\
&-\frac{N-1}{2}\Big(V'(\phi)/\phi\Big)\int_{-\pi/a}^{+\pi/a}\frac{d^4k}{(2\pi)^4}\frac{1}{\hat{k}^2} \\
&+\frac{N-1}{4}\Big(V'(\phi)/\phi\Big)^2\int_{-\pi/a}^{+\pi/a}\frac{d^4k}{(2\pi)^4}\frac{1}{(\hat{k}^2+\mu^2)^2} .
\end{aligned}
\tag{6}
$$

The vacuum expectation value v is determined by the minimum of the effective potential, $U'_\Omega(v) = 0$. In one-loop approximation we find

$$
\begin{aligned}
v = \phi_{tree} \cdot \Big[&1 - \frac{3}{2\phi_{tree}^2}\Big(\frac{1}{\Omega}\sum_{\hat{k}\neq 0}\frac{1}{\hat{k}^2+2m^2} - J_1(0) - 2m^2 J_2(\mu)\Big) \\
&- \frac{N-1}{2\phi_{tree}^2}\Big(\frac{1}{\Omega}\sum_{\hat{k}\neq 0}\frac{1}{\hat{k}^2} - J_1(0)\Big)\Big] ,
\end{aligned}
\tag{7}
$$

where $\phi_{tree} = \sqrt{m(\mu)^2/4\lambda(\mu)}$. The notation

$$
J_p(\mu) = \int_{-\pi/a}^{+\pi/a}\frac{d^4k}{(2\pi)^4}\frac{1}{(\hat{k}^2+\mu^2)^p}
\tag{8}
$$

was used in Eq. (7). The formula in Eq. (7) is the basis of a finite size scaling analysis for the vacuum expectation value of the scalar field.[2]

To calculate the renormalized mass m_R in terms of finite tree level parameters m and λ, we will need the second derivative of the effective potential at the minimum with $m_R^2 = Z_\Phi \cdot U''_\Omega(v)$, in the one-component Φ^4 model. For general O(N) we find

$$
\begin{aligned}
U''_\Omega(v) = 2m^2 \cdot \Big[&1 - \frac{12\lambda}{m^2}\Big(\frac{1}{\Omega}\sum_{\hat{k}\neq 0}\Big[\frac{1}{\hat{k}^2+2m^2} + \frac{3m^2}{(\hat{k}^2+2m^2)^2}\Big] - J_1(0) - m^2 J_2(\mu)\Big) \\
&- \frac{4(N-1)\lambda}{m^2}\Big(\frac{1}{\Omega}\sum_{\hat{k}\neq 0}\Big[\frac{1}{\hat{k}^2} + \frac{m^2}{\hat{k}^4}\Big] - J_1(0) - m^2 J_2(\mu)\Big)\Big] ,
\end{aligned}
\tag{9}
$$

where the last term of Eq. (9), which is non-vanishing only for $N \geq 2$, has infrared problems in the infinite volume limit. It will be shown that the finite lattice volume regulates the infrared problem and Eq. (9) can be used in a useful way.[2]

Higher derivatives of the effective potential will be related to the three-point, four-point, and higher couplings of the theory. In one-loop approximation the wavefunction renormalization constant Z_Φ is one in the symmetric phase.

The loop expansion of the effective potential will be compared with our Monte Carlo data. For accurate predictions, we will perform a leading logarithmic summation of the loop expansion by using standard renormalization group (RG) techniques.

1.2. Renormalization group improved effective potential

The renormalized effective potential $U_\Omega(\phi)$, which does not depend on the renormalization scale μ, satisfies the renormalization group equation,

$$\left[-\frac{\partial}{\partial t} + \beta_\lambda \frac{\partial}{\partial \lambda} + \beta_m \cdot m^2 \frac{\partial}{\partial m^2} \right] U_\Omega(\phi) = 0 \ , \tag{10}$$

where

$$t = \ln \left(\frac{\phi}{\mu} \right) ,$$

$$\beta_\lambda = \frac{N+8}{2\pi^2} \lambda^2 , \tag{11}$$

$$\beta_m = \frac{N+2}{2\pi^2} \lambda ,$$

in one-loop approximation (in higher order there would be an additional term in the RG equation).

The solution to Eqs. (10) and (11) has the simple tree level form, but with a mass and coupling constant dependent on ϕ,

$$U_\Omega(\phi) = -\frac{m^2(t)}{2} \phi^2 + \lambda(t) \phi^4 ,$$

$$\lambda(t) = \frac{\lambda(0)}{1 - \frac{N+8}{2\pi^2} \lambda(0) t} , \tag{12}$$

$$m^2(t) = m^2(0) \left[1 - \lambda(0) \frac{N+8}{2\pi^2} t \right]^{-\frac{N+2}{N+8}} .$$

When $U_\Omega(\phi)$ of Eq. (12) is expanded in the variable $\lambda(0) \ln \frac{\phi}{\mu}$, in first non-trivial order we will recover the leading logarithmic term of the one-loop effective potential.

However, with the RG technique we summed the leading logs to arbitrary order in the loop expansion. First, we will apply the results to the one-component Φ^4 model.

2. ONE-COMPONENT Φ^4 THEORY

The one-component Φ^4 model, with the lattice action of Eq. (1) but without internal O(4) indices, is a warm-up exercise to our more realistic Monte Carlo studies of the O(4) model. Since the model only has a discrete $\Phi \rightarrow -\Phi$ symmetry, Goldstone particles are not associated with the spontaneously broken phase which makes the calculations easier.

2.1. Triviality bound

In this simple one-component scalar theory we can extract renormalized physical parameters from the effective potential as it was discussed earlier.[1] The critical line in the parameter space of the bare coupling and bare mass is determined from the effective potential. For independent information we also used the histograms of unconstrained sampling to determine the probability distribution of the average "magnetization" over a large number of Monte Carlo configurations. Table 1 is a summary of the three critical points where we have done large simulations in the model.

λ	m_c^2	lattice sizes
25	24.55(5)	8^4, 12^4
100	74.4(1)	8^4
∞	$\kappa = 0.1495(2)$	8^4, 12^4, 16^4, 18^4

Table 1   The critical values of m^2 are given in the Φ^4 model for several values of λ. At infinite bare coupling the mass parameter is replaced by κ which is the nearest neighbor coupling between two lattice sites (our definition of κ differs from the notation of ref. 3 by a factor of two).

For fixed bare coupling λ we calculated the effective potential at several values of the bare mass m^2 on the two sides of the phase transition line. The two-point function in momentum space was also simulated in unconstrained Monte Carlo runs at every value of bare m^2 for an independent determination of the renormalized mass m_R and the wavefunction renormalization constant Z_Φ. The extraction of these parameters from momentum space propagator measurements will be discussed in the more complicated $O(4)$ model. The technique for the Φ^4 model was discussed elsewhere.[1,4,5]

From the analysis of the momentum space propagator and the effective potential we determined the renormalized coupling constant λ_R in the symmetric phase, the renormalized mass m_R and the wavefunction renormalization constant Z_Φ in both phases, and the vacuum expectation value $\langle \Phi_R \rangle$ of the renormalized field operator in the broken symmetry phase.

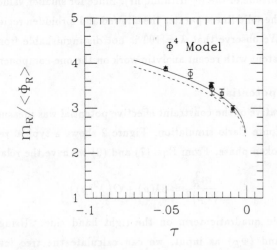

Figure 1. The triviality upper bound on the Higgs mass is illustrated with new data of the $m_R/\langle \Phi_R \rangle$ ratio. The list of the symbols is given in Table 2 for the identification of data points according to coupling and lattice size. The dashed line is our earlier result (refs. 4,5) which was obtained at $\lambda = 25$.

Our new data on the triviality mass bound of the heavy scalar particle is depicted in Figure 1 as a function of the variable $\tau = 1 - m^2/m_c^2$. At infinite bare coupling λ we fitted the ratio $m_R/\langle \Phi_R \rangle$ to a logarithmic function of τ as required by the trivial Gaussian fixed point for small values of τ. The solid line is a fit to the form $m_R/\langle \Phi_R \rangle = const \left| \ln |\tau| \right|^{-\frac{1}{2}}$, with the fitted value of 6.6(1) for the constant. The dashed line is our earlier result[4,5] at $\lambda = 25$, approximately six percent lower than new results at ∞ coupling. A detailed finite size analysis is required to confirm

symbol	λ	lattice size
open circle	100	8^4
full triangle	∞	12^4
open box	∞	16^4
full box	∞	18^4

Table 2 Symbols are shown for the Φ^4 model data at two values of the bare coupling λ, for different lattice sizes.

this difference (finite size effects may have a slightly bigger effect than the statistical error in the amplitude of the logarithmic fit). Since for smaller values of λ the ratio data are lower, the solid line separates the allowed and forbidden regions of $m_R/\langle \Phi_R \rangle$ in the theory. We observe that $\lambda = 100$ is not distinguishable from $\lambda = \infty$. Our results are consistent with recent analytic work on the one-component Φ^4 model.[6,7]

2.2. Effective potential

The derivative of the constraint effective potential was measured in a Fourier space Hybrid Monte Carlo simulation. Figure 2 shows a typical result for the Φ^4 theory in the broken phase. From Eqs. (7) and (9) we have the relation

$$\frac{m_R^2}{\langle \Phi_R \rangle^2} = 8\lambda(\mu) + O\left(\lambda^2(\mu)\right) , \tag{13}$$

with a calculable quadratic term on the right hand side. Using the measured values of m_R and $\langle \Phi_R \rangle$ as input, we can calculate the tree level potential in terms of $\lambda(\mu)$ and $m(\mu)$ which is shown as the dashed line in Figure 2. The one-loop form of Eq. (3) is depicted by the dashed-dotted line. The RG improved constraint effective potential $U_\Omega(\phi)$ is plotted as the solid line in Figure 2. To obtain reasonable agreement between our Monte Carlo result and renormalized perturbation theory, renormalization group improvement was necessary which explains our earlier struggle to match Monte Carlo data to one-loop formulae in similar parameter range.

For internal consistency, we also determined the renormalized three-point coupling λ_{3R} from the measured zero momentum three-point function in unconstrained Monte Carlo runs in the limit of infinite bare coupling. Figure 3 is a summary of the results.

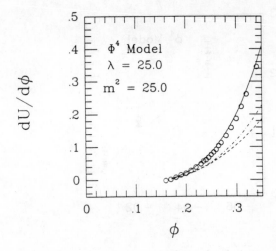

Figure 2. The derivative of the effective potential in the Φ^4 model is shown in the broken phase at lattice size 8^4. The renormalized parameters $m_R = 0.53(2)$ and $\langle \Phi_R \rangle = 0.164(5)$ were also determined. The solid line is the RG improved constraint effective potential.

3. O(4) MODEL

The generalization of the above techniques to the four-component scalar field is straightforward. Some care has to be exercised in extrapolating to the infinite volume limit, since the massless Goldstone modes in the broken symmetry phase generate infrared singularities at zero momentum. The finite lattice volume, however, acts as a good infrared regulator and physical results can be extracted as we will describe it now.

3.1. The Higgs propagator

In the broken symmetry phase, due to the presence of massless Goldstone modes as intermediate particles, the longitudinal Higgs propagator $\tilde{G}(p^2)$ has a branch cut starting at zero momentum, in the infinite volume limit. The massive Higgs particle appears as a pole in the complex p^2 plane. The real part of the pole defines the mass of the Higgs particle and the imaginary part of the pole defines the width of the particle. For small renormalized coupling constant, the Higgs particle is a well defined object with a small width and sharp peak in the spectral function

224

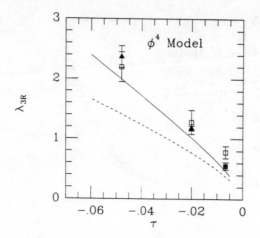

Figure 3. The renormalized coupling constant λ_{3R} is shown in the broken symmetry phase of the Φ^4 model. The list of the data symbols is given in Table 2. The dashed line is the tree-level value of λ_{3R} using the ratio $m_R/\langle\Phi_R\rangle$ as input. The solid line is a one-loop calculation using input results from Figure 1, as explained in the text.

of the propagator. Even for a 600 GeV Higgs mass the width is less than twenty percent of the mass, and the pole is close to the real axis.[8]

The simplest infrared diagram which contributes to the branch cut is shown in Figure 4. To avoid the infrared singularity in the infinite volume limit, the definitions of $Z_\Phi(\mu^2)$ and the renormalized mass $m_R(\mu^2)$ are taken at some nonzero Euclidean point μ^2:

$$\partial \tilde{G}(p^2)^{-1}/\partial p^2 \Big|_{p^2=\mu^2} = Z_\Phi^{-1}(\mu^2) \,,$$
$$\tilde{G}^{-1}(\mu^2) = Z_\Phi^{-1}(\mu^2)\left[\mu^2 + m_R^2(\mu^2)\right] \,. \tag{14}$$

The real part of the complex pole in the longitudinal Higgs propagator defines the physical mass m_H which differs from $m_R(\mu^2)$ by a finite and calculable small amount in perturbation theory. The momentum space propagator has the form

$$\tilde{G}(\hat{p}^2)^{-1} = \hat{p}^2 + m^2(\mu) + \left(\sum(\hat{p}^2) - \sum(\hat{p}^2 = \mu^2)\right) \,, \tag{15}$$

where $\sum(\hat{p}^2)$ designates the self-energy part of the propagator, and the precise definition of the tree-level $m(\mu)$ is given by the counterterms. The contribution of

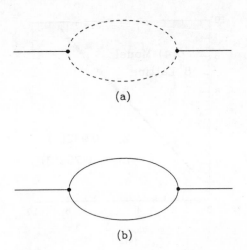

(a)

(b)

Figure 4. Diagrams (a) and (b) are the leading contributions in λ to the Higgs propagator, in the broken symmetry phase. Solid lines represent the massive Higgs particle and dashed lines designate the three Goldstone particles with a summation included in the loop. Diagram (a) has an infrared divergence in the $\Omega \to \infty$ limit.

Figure 4a to the self-energy part $\sum(\hat{p}^2)$ is given by the finite lattice sum,

$$\sum(\hat{p}^2) = -8(N-1)\lambda m^2 \frac{1}{\Omega} \sum_{\substack{\hat{q} \neq 0 \\ \hat{q} \neq \hat{p}}} \frac{1}{\hat{q}^2 \cdot (\hat{q} - \hat{p})^2} \, , \tag{16}$$

where the factor $N - 1$ is the Goldstone particle count in the loop. There is also an infrared finite contribution from Figure 4b.

Figure 5 shows the inverse of the longitudinal Higgs propagator on a finite lattice as plotted against the inverse of the free and massless longitudinal lattice propagator. The solid straight line is a fit to our data. With our choice of the bare parameters in the run, the straight line fit corresponds to a free particle of mass $m_R = 0.753(4)$ (the error in the fit is somewhat underestimated as the scatter of repeated runs indicates). The slope of the straight line is Z_Φ which is close to one in the fit.

The approximate free particle behavior of the full propagator is well understood in perturbation theory on the finite lattice. The dashed line we plot is $\tilde{G}(\hat{p}^2)^{-1}$ from Eq. (15) with the self-energy contributions of Figure 4 on the finite lattice. For large range of lattice momenta, there is very small deviation from the straight line. The $\hat{p}^4$ terms, or higher order terms, in the self-energy part are very

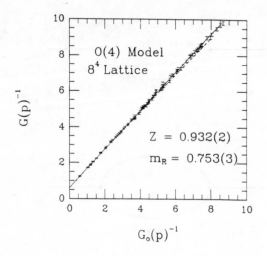

Figure 5. The inverse momentum space propagator in the O(4) model

small in perturbation theory, and they are practically not visible within the accuracy of the Monte Carlo simulation.

The logarithmic effect of the infrared term, which would drive the dashed line in Figure 5 to zero at $\hat{p}^2 = 0$ on an infinite lattice, could only be seen at very low values of the lattice momentum. Small lattice momenta in that infrared range are simply not present in our finite lattice simulations (they could only be reached on very large lattices). For lattice propagator data along an approximately straight line for finite lattice momenta, $m_R^2(\mu)$ can be determined from the intercept of the extended straight line at zero lattice momentum, according to our definition in Eq. (14). Numerically, we find only a few percent difference between $m_R^2(\mu^2)$ and m_H^2 in our analysis.[2]

3.2. Effective potential

In the infinite volume limit the constraint O(4) effective potential, at its minimum, has infrared singularities diagram by diagram in perturbation theory. On the finite lattice, however, the theory is infrared regulated. Infrared regularization works very similarly to the technique we presented for the longitudinal propagator.

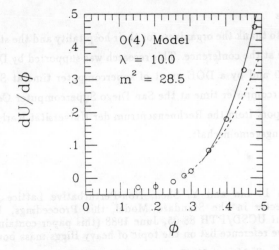

Figure 6. *The derivative of the effective potential in the O(4) model is shown in the broken phase at lattice size 8^4. The dashed line is the tree level potential and the solid line is the result from RG improved loop expansion.*

4. CONCLUSION

We will now summarize the main results of this report. First, we have shown that RG improved loop expansion works well for the constraint effective potential in the critical regions of the Φ^4 and $O(4)$ models. Second, we explained the approximate straight line form of the momentum space propagators in our Monte Carlo simulations by applying renormalized perturbation theory to the Euclidean Green's functions on the finite lattice. Third, we have shown that the infrared singular behavior of the ∞ volume propagator and constraint effective potential is not visible on lattice sizes where simulations are feasible today. The infrared problem of the infinite lattice did not present any serious practical problem in our Monte Carlo calculation of physical parameters. We have to conclude that no evidence has been seen in our work for a strongly interacting Higgs sector below the Higgs mass bound which we established earlier.

228

Acknowledgements

We would like to thank the organizers for their hospitality and the stimulating workshop atmosphere at the conference. This research was supported by DOE grant DE-AT03-81-ER40029 and by a DOE grant of supercomputer time at SCRI, and an NSF grant of supercomputer time at the San Diego Supercomputer Center. We also acknowledge support from the Rechnenzentrum der Universität Karlsruhe and the Deutsche Forschungsgemeinschaft.

REFERENCES

1. J. Kuti with L. Lin and Y. Shen, Non-Perturbative Lattice Study of the Higgs Sector in the Standard Model, this Proceedings, U.C. San Diego preprint UCSD/PTH 88-05, June 1988 (this paper contains a more comprehensive reference list on the topic of heavy Higgs mass bound).

2. J. Kuti, L. Lin, S. Meyer, Y. Shen, Non-Perturbative Lattice Study of the O(4) Model, U.C. San Diego preprint UCSD/PTH 88-08, July 1987.

3. M. Lüscher and P. Weisz, Nucl. Phys., 290[FS20], (1987) 25.

4. J. Kuti and Y. Shen, Phys. Rev. Lett., 60, (1988) 85.

5. J. Kuti, L. Lin and Y. Shen, Nucl. Phys. (Proc. Suppl.), B4, (1988) 397.

6. P. Weisz, Scaling Laws and Triviality Bounds in the Lattice N Component ϕ^4 Theory, this Proceedings.

7. M. Lüscher and P. Weisz, Nucl. Phys., B295[FS21], (1987) 65.

8. L. Lin with J. Kuti and Y. Shen, How Far is N=4 from N=∞?, this Proceedings, U.C. San Diego preprint UCSD/PTH 88-07, June 1988.

YUKAWA COUPLINGS ON THE LATTICE

J. Shigemitsu

Physics Department
The Ohio State University
Columbus, Ohio 43210
USA

ABSTRACT

Strong Yukawa couplings between fermions and scalars could have significant feedback onto the Higgs sector of the Standard Model. Some preliminary attempts to study this coupling on the lattice are discussed.

1. INTRODUCTION

Recently there has been considerable interest in fermion mass generation in theories that undergo spontaneous symmetry breaking (SSB).[1-6] We currently know of two mechanisms for creating fermion masses dynamically: 1. chiral symmetry breaking (χSB) and 2. through Yukawa couplings to scalars when the scalar sector is in the spontaneously broken phase. The first mechanism has been studied extensively on the lattice. (χSB) is a very common phenomenon in vectorlike gauge theories and takes place whenever the interfermion coupling becomes sufficiently large. Except for some interesting large N studies, [4] the second mechanism (Yukawa couplings) has been investigated almost exclusively within perturbation theory. The main reason for this is that for all known quarks and leptons, Yukawa couplings are weak. However, with the lower bound on the top quark mass getting higher and higher and the possibility that one may soon be contemplating a fourth generation, it may be time to start thinking more about nonperturbative approaches to Yukawa couplings.

At this workshop we have heard about several very interesting studies of λ_R, the renormalized λ_φ^4 coupling, in the $O(4)$ model. At the present time there is little evidence for a strongly coupled effective scalar theory in the scaling region of the lattice model. Several groups have converted this information into upper bounds on the Standard Model Higgs particle mass.[7] When moving on to another important sector of the Standard Model, namely the fermionic sector, and focussing on the Yukawa coupling y, one wishes ultimately to be able to address similar issues.

- Is there a bound on the renormalized coupling y and hence possibly also on fermion masses?

- What is the feedback of fermions onto the scalar sector and onto the behavior of λ_R?

It is unclear whether lattices can shed light on the above questions. In the case of λ_R a vast literature existed already on the "triviality" problem. Here almost nothing is known. Furthermore in order to make any progress one must gain control over lattice fermions.

In this talk I wish to describe two attempts to investigate Yukawa couplings on the lattice. One project discusses changes in the lattice phase diagram of a gauge-Higgs system due to fermions interacting directly with both the gauge and the scalar fields. The second project is aimed at developing tools to investigate dynamical fermions in models with Yukawa couplings.

2. SU(2) GAUGE-HIGGS SYSTEM WITH FERMIONS

The degrees of freedom in this model are:

$$\text{Scalars}: \phi = \begin{pmatrix} \varphi_1 \\ \varphi_2 \end{pmatrix} \text{ with } |\varphi_1|^2 + |\varphi_2|^2 = 1$$

they can also be introduced as matrices

$$\Omega \equiv \begin{pmatrix} \varphi_1 & -\varphi_2^* \\ \varphi_2 & \varphi_1^* \end{pmatrix}$$

Fermions: SU(2) doublet fields $\chi, \bar{\chi}$

           SU(2) singlet fields $\xi_f, \bar{\xi}_f$

$$(f = 12)$$

Gauge fields: U

The action is:

$$S = S_G + S_H + S_F + S_Y \tag{1}$$

$$S_G = \beta_g \sum_{\text{plaq.}} \left[1 - \frac{1}{2} tr U_p \right] \tag{2}$$

$$S_H = -\beta_H \sum_{n,\mu} tr \left(\Omega^\dagger(n) U_\mu(n) \Omega(n+\mu) \right) \tag{3}$$

$$S_F = \sum_{n,\mu} \bar{\chi}(n)\eta_\mu(n)\frac{1}{2} \left[U_\mu(n)\chi(n+\mu) - U_\mu^\dagger(n-\mu)\chi(n-\mu) \right] \tag{4}$$

$$+ \sum_{n,\mu} \sum_{f=1}^{2} \bar{\xi}_f(n)\eta_\mu(n)\frac{1}{2} \left[\xi_f(n+\mu) - \xi_f(n-\mu) \right]$$

$[\eta_\mu(n)$: usual staggered fermion phases]

$$S_Y = \left[y_1 \sum_n \bar{\chi}(n)\phi(n)\xi_1(n) + y_2 \sum_n \bar{\chi}(n)(-i\tau_2\phi^*(n))\xi_2(n) \right] + [h.c.] \tag{5}$$

For $y_1 = y_2 \equiv y$

$$S_Y \to y \sum_n [\bar{\chi}\Omega\xi + h.c.]$$

$$\left(\xi \equiv \begin{pmatrix} \xi_1 \\ \xi_2 \end{pmatrix} \right) \tag{6}$$

The action (1) has a 3-dimensional parameter space (β_g, β_H, y) associated with it. The $y = 0$ surface has been the focus of some attention in the past few years.[1,2] In this limit the ξ-fields decouple completely and all of the interesting fermionic dynamics resides in the χ-fields. Both analytic and numerical work have shown that the $y = 0$ plane is divided into two phases characterized by whether $\langle\bar{\chi}\chi\rangle = 0$ or not. Another limiting plane that is easy to analyze is the $y \to \infty$ surface. If one rescales

$$x \to \frac{1}{\sqrt{y}}\chi, \quad \xi \to \frac{1}{\sqrt{y}}\xi$$

then,

$$S_F + S_Y \to \frac{1}{y}\{\text{Hopping terms}\} + \sum_n [\bar{\chi}\Omega\xi + h.c.]$$

$$\equiv \frac{1}{y}\{\text{Hopping terms}\} + \bar{\psi}M^{(0)}\psi \tag{7}$$

Since $det\left(M^{(0)}\right) = 1$ the fermions decouple at $y \to \infty$ and one has the same phase diagram as in the pure gauge Higgs system. One can evaluate y^{-1} corrections to the zeroth order partition function and finds that this brings in the following addition to the action

$$-\frac{1}{(2y)^2}\left[tr\left(\Omega^\dagger(n)U_\mu(n)\Omega(n+\mu)\right)+h.c.\right] \tag{8}$$

which is of the same form as S_H. In other words, the y^{-2} corrections bring about a shift

$$\beta_H \rightarrow \beta_H + \frac{1}{wy^2}. \tag{9}$$

The other nontrivial surfaces about which something can be said are the $\beta_g = 0$ and the $\beta_0 \rightarrow \infty$ planes. Fig. 1 summarizes the different limiting surfaces in the (β_g, β_H, y) parameter space. We are currently working towards getting more information about the interior of the cube. It is clear that one will end up with two distinct phases characterized by $\langle \bar{\chi}\chi \rangle$. At this point it may be useful to remind ourselves of the symmetries of the lattice model so that one can interpret the implication of $\langle \bar{\chi}\chi \rangle \neq 0$. What continuum symmetries are being broken?

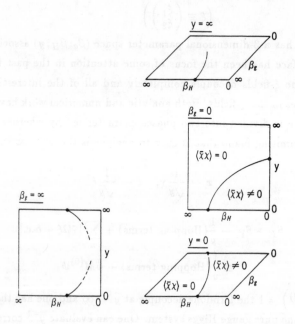

Figure 1. Phase diagrams on several surfaces in the (β_g, β_H, y) parameter space.

Before introducing fermions the scalar sector is symmetric under

$$\Omega \to G_L(\text{local})\Omega G_R^{-1}(\text{global})$$
$$(G_{L,R} \in SU(2))$$

(10)

The Yukawa term $\bar{\chi}\Omega\xi$ also respects (10) provided

$$\chi \to G_L(\text{local})\chi$$
$$\xi \to G_R(\text{global})\xi.$$

(11)

In addition one has the $U_{\text{even}}(1) \otimes U_{\text{odd}}(1)$ global symmetries of staggered fermions
For $y = 0$ these are:

$$n = \text{even} \quad \begin{matrix} \chi \to e^{i\alpha_e}\chi & \xi \to e^{i\beta_e}\xi \\ \chi \to e^{-i\alpha_0}\bar{\chi} & \xi \to e^{-i\beta_0}\bar{\xi} \end{matrix}$$

$$n = \text{odd} \quad \begin{matrix} \chi \to e^{i\alpha_0}\chi & \xi \to e^{i\beta_0}\xi \\ \chi \to e^{-i\alpha_e}\bar{\chi} & \xi \to e^{-i\beta_e}\bar{\xi} \end{matrix}$$

(12)

The Yukawa term requires

$$\begin{cases} \alpha_0 = \beta_e \\ \alpha_e = \beta_0 \end{cases}$$

(13)

and one is still left with 2 independent $U(1)$'s.

In order to interpret (13) recall that

$$\chi - \text{fields} \to 4 \text{ flavors of local } SU(2) \text{ doublets, } \psi$$

$$\chi - \text{fields} \to \text{another set of 4 global } SU(2) \text{ doublets, } \bar{\psi}.$$

In this language one has

$$I \text{ for } \alpha_0 = \alpha_e \qquad \psi \to e^{i\alpha}\psi, \bar{\psi} \to e^{i\alpha}\bar{\psi}$$

(14)

$$II \text{ for } \alpha_e = -\alpha_0 \qquad \psi \to e^{i\alpha\gamma_5 T}\psi, \bar{\psi} \to e^{-i\alpha\gamma_5 T}\bar{\psi}$$

(15)

where $T = 4 \times 4$ flavor matrix.

$\langle\bar{\chi}\chi\rangle \neq 0$ implies that the symmetry II is spontaneously broken.

It is amusing to remind ourselves that Yukawa coupling generated fermion masses become the "bare" quark masses of QCD. So,

$$\psi_{QCD} = \left[P_L\left(\Omega^\dagger \psi\right) + P_R\bar{\psi}\right]$$
$$\left(P_{L,R} = \frac{1}{2}(1 \pm \gamma_5)\right).$$

(16)

Since our model is vectorlike, one has in addition to (16)

$$\psi'_{QCD} = \left[P_R \left(\Omega^\dagger \psi \right) + P_L \tilde{\psi} \right]. \tag{17}$$

For the "QCD spinors" the symmetry II corresponds to

$$\begin{aligned}
\psi_{QCD} &\to e^{-iaT} \psi_{QCD} \\
\psi'_{QCD} &\to e^{-iaT} \psi'_{QCD}.
\end{aligned} \tag{18}$$

In other words II is a "vector" symmetry in the QCD basis.

There is a way to introduce fermions such that the symmetry II is absent. Namely one can work with "reduced staggered fermions."[8]

$$\bar{\xi}, \chi - \text{fields only on } n = \text{odd sites.}$$

$$\xi, \bar{\chi} - \text{fields only on } n = \text{even sites.}$$

The fermionic part of the action is changed to

$$\begin{aligned}
S'_F = &\sum_{n=\text{even}} \sum_\mu \bar{\chi}(n) \eta_\mu(n) \frac{1}{2} \Big\{ U_\mu(n) \chi(n-\mu) - U_\mu^\dagger(n-\mu) \chi(n-\mu) \Big\} \\
&+ \sum_{n=\text{odd}} \sum_\mu \sum_{f=1} \bar{\xi}_f(n) \eta_\mu(n) \frac{1}{2} \Big\{ \xi_f(n+\mu) - \xi_f(n-\mu) \Big\}
\end{aligned} \tag{19}$$

$$S'_Y = y \sum_{n=\text{even}} (\bar{\chi} \Omega \xi) + y \sum_{n=\text{odd}} (\bar{\xi} \Omega \chi). \tag{20}$$

This fermionic action has only one global $U(1)$ symmetry. Furthermore, single site mass terms $\bar{\chi}\chi$ or $\bar{\xi}\xi$ are not possible. On the other hand, for the $SU(2)$ gauge Higgs systems $\langle \chi \varepsilon \chi \rangle$ and $\langle \bar{\chi} \varepsilon \bar{\chi} \rangle$, associated with it $\left(\varepsilon \equiv i\tau_2 \right)$ condensates are still possible. Mean Field analyses at $\beta_g = 0$ show that there is indeed a phase in which a condensate forms. In fact, within the $\beta_g = 0$, mean field approximation the $SU(2)$ gauge Higgs system with reduced fermions maps onto an Abelian $U(1)$ Higgs model with nonreduced fermions. This latter model has been studied by several groups.[2] Using their results one can conclude that, at least for small β_g, the phase diagram of the $SU(2)$ model is not altered in any qualitative (or even quantitative) way, by going from conventional to reduced staggered fermions. When one considers that in the continuum limit the only difference between reduced and nonreduced fermions is that the former describes 2 and the latter 4 flavors, we find this reassuring. We note, however, that the gauge group $SU(2)$ may be very special in this respect.

3. THE $\lambda\varphi^4$ THEORY WITH YUKAWA COUPLINGS TO STAGGERED FERMIONS

This is the simplest model in which to study the feedback of dynamical fermions on the scalar sector. The action is given by

$$S = S_B + S_F + S_Y \tag{21}$$

$$S_B = \sum_n \left\{ 4\varphi^2(n) - \sum_\mu \varphi(n)\varphi(n+\mu) \right\} + \frac{m^2}{2}\sum_n \varphi^2(n) + \frac{\lambda}{4}\sum_n \varphi^4(n) \tag{22}$$

$$S_F = \frac{1}{2}\sum_n \sum_\mu \sum_{j=1}^2 \bar{\chi}_j(n)\eta_\mu(n)\left[\chi_j(n+\mu) - \chi_j(n-\mu)\right] \tag{23}$$

For the Yukawa coupling term S_y we experimented with 2 versions

$$S_Y^1 = y\sum_n \bar{\chi}(n)\chi(n)\varphi(n) \tag{24}$$

or

$$S_Y^2 = y\sum_n \varphi(n)\left(\frac{1}{16}\sum_{\text{hyperc.}}\bar{\chi}\chi\right) \equiv y\sum_n \bar{\chi}(n)\chi(n)\Phi(n)$$

$$\left(\Phi(n) \equiv \frac{1}{16}\sum_{\text{hyperc.}}\varphi(n)\right). \tag{25}$$

We find that the lattice phase diagram and the stability of numerical algorithms depend sensitively on the choice of S_Y. This shows up most clearly in the large y limit, where one can carry out the integration over the χ-fields. One ends up with,

$$S_{\text{eff}}^1 = S_B - \sum_n \ln\left(\varphi(n)\right)^2 \tag{26}$$

or

$$S_{\text{eff}}^2 = S_B - \sum_n \ln\left(\Phi(n)\right)^2. \tag{27}$$

In the Ising limit the second term in (26) vanishes, whereas (27) requires

$$\Phi = \frac{1}{16}\sum_{\text{hyperc.}}\varphi \neq 0. \tag{28}$$

Even away from this limit one expects S_Y^2 to reenforce spontaneous symmetry breaking ($\langle\varphi\rangle \neq 0$) much more strongly than S_Y^1. The different behavior of S_Y^1 and

236

S_Y^2 is reminiscent of the situation in attempts to formulate the Gross Neveu Model on a 2D Euclidean lattice.[9] There too, different lattice actions led to different phase diagrams. What this tells us is that one must be extremely careful to distinguish between lattice artifacts and results relevant to continuum physics.

We have developed hybrid molecular dynamic codes to analyse the full action, including dynamical fermions. Most of our data were obtained using S_Y^2. With S_Y^1 our numerical methods occasionally seemed to run into stability problems having to do with zero modes. Fig. 2 shows $\langle \varphi \rangle$ as a function of y at $\lambda = 1$ and for several m^2 values. This information can be converted into a phase diagram in the (m^2, y) plane. In Fig. 3 we plot the curves separating the spontaneously broken from the symmetric regions for $\lambda = 1$ and $\lambda = 100$.

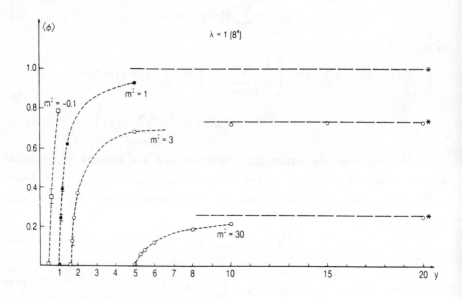

Figure 2. $\langle \varphi \rangle$ versus y at $\lambda = 1$ and for several m^2 values. The Yukawa term S_Y^2 was used for these simulations. The stars to the right show the Monte Carlo results for the infinite y effective action S_{eff}^2 of Eq. (27).

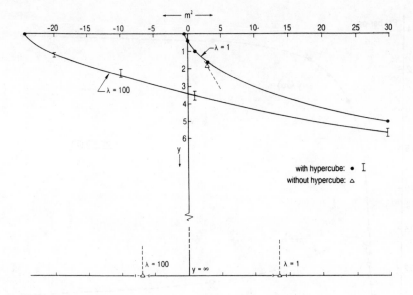

Figure 3. The phase diagram in the (y, m^2) parameter space for $\lambda = 1$ and $\lambda = 100$. The lines separate the symmetric (large m^2) from the broken (small m^2) regions.

The model (21) has a discrete chiral symmetry,

$$\chi(n) \to i(-1)^{n_0 + n_2} \chi(n + \xi)$$
$$\bar{\chi}(n) \to i(-1)^{n_0 + n_2} \bar{\chi}(n + \xi) \tag{29}$$
$$\varphi(n) \to -\varphi(n + \xi).$$

$$(\text{where } \xi = \hat{e}_0 + \hat{e}_1 + \hat{e}_2 + \hat{e}_3)$$

(29) is broken in the $\langle \varphi \rangle \neq 0$ phase, so one also expects $\langle \bar{\chi}\chi \rangle \neq 0$ there. Fig. 4 shows $\langle \bar{\chi}\chi \rangle$ for $\lambda = 1$, $m^2 = 3$ as a function of y. Just as with $\langle \varphi \rangle$ in Fig. 2 one sees this order parameter becoming non-zero at $y \sim 1.6 - 1.7$ and rapidly taking on the large y behavior ($\langle \bar{\chi}\chi \rangle \propto 1/y$).

Ref. [6] describes the dynamical fermion algorithms used in more detail. We believe that with the hypercube Yukawa term, S_Y^2, our version of a hybrid molecular dynamic algorithm works well. The next step is to devise methods to measure y_R,

238

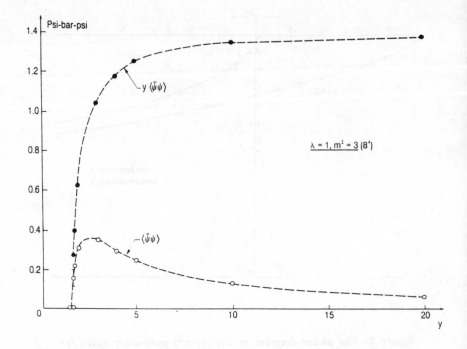

Figure 4. $\langle \bar{\chi}\chi \rangle$ *and* $y \langle \bar{\chi}\chi \rangle$ *versus y at* $\lambda = 1$ *and* $m^2 = 3$.

the renormalized Yukawa coupling, and follow its behavior as one tries to take a continuum limit. To this end, one could attempt a direct evaluation of the scalar-fermion-antifermion three point function. Methods used in QCD to extract the $\rho\pi\pi$ coupling may be applicable here.[10] Other interesting physical quantities would be ratios such as $M_{\text{scalar}}/M_{\text{fermion}}$ and $M_{\text{fermion}}/\langle \varphi_r \rangle$. Work along these lines is underway.

Acknowledgements

This research is supported in part by grants from the U.S. Department of Energy DE-AC02ER01545 and DE-AC02-76ER03069 and by a CRAY Inc. Grant. The numerical work was carried out on the CRAY X-MP/24 of the Ohio Supercomputer Center. Finally, I wish to thank and congratulate the organizers of the Lattice Higgs Workshop for a very stimulating meeting.

REFERENCES

1. I.-H. Lee and J. Shigemitsu, Phys. Lett., 178B, (1986) 93.
   A. De and J. Shigemitsu, OSU Preprint, DOE-ER-01545-398, to appear in Nucl. Phys. B.

2. I.-H. Lee and R.E. Shrock, Phys. Rev. Lett., 59, (1987) 14.
   I.-H. Lee and R.E. Shrock, Nucl. Phys., B290, (1987) 275.
   I.-H. Lee and R.E. Shrock, Phys. Lett., 196B, (1987) 82.
   I.-H. Lee and R.E. Shrock, Phys. Lett., 199B, (1987) 541.
   I.-H. Lee and R.E. Shrock, Phys. Lett., 201B, (1988) 497.
   E. Dagotto and J. Kogut, Ill-(TH)-88-9.

3. P. Swift, phys. Lett., 145B, (1984) 256.
   J. Smit, Acta Physica Polonica, B17, (1986) 531.
   S. J. Hand and D. B. Carpenter, Nucl. Phys., B266, (1986) 285.
   J. Shigemitsu, Phys. Lett., 189B, (1987) 164 OSU Preprint, DOE-ER-01545-397.
   I. Montvay, DESY-87-077; DESY-87-110; DESY-88-028.

4. M. B. Einhorn and G. J. Goldberg, Phys. Rev. Lett., 57, (1986) 2115.

5. A. Hasenfratz and T. Neuhaus, FSU-SCRI Preprint, May 1988.
   D. Stephenson and A. Thornton, Edinburgh Preprint 88/436.

6. J. Polonyi and J. Shigemitsu, OSU Preprint DOE-ER-01545-403.

7. M. Lüscher and P. Weisz, Nucl. Phys., B290[FS20], (1987) 25.
   M. Lüscher and P. Weisz, Nucl. Phys., B295[FS21], (1988) 65.
   J. Kuti and Y. Shen, Phys. Rev. Lett., 60, (1988) 85.
   A. Hasenfratz, T. Neuhaus, K. Jansen, Y. Yoneyama, and C.B. Lang, Phys. Lett., 199B, (1987) 531.

8. H.S. Sharatchandra, H.J. Thun, and P. Weisz, Nucl. Phys., B192, (1982) 205.
   T. Banks and A. Zaks, Nucl. Phys., B206, (1982) 23.
   C. Van Den Doel and J. Smit, Nucl. Phys., B228, (1983) 122.

9. Y. Cohen, S. Elitzur and E. Rabinovici, Phys. Lett., 104B, (1981) 289.
   Y. Cohen, S. Elitzur and E. Rabinovici, Nucl. Phys., B220[FS8], (1983) 102.
   I. Affleck, Phys. Lett, 109B, (1982) 307.

10. S. Gottlieb, P.B. MacKenzie, H.B. Thacker, and D. Weingarten, Phys. Lett., 134B, (1984) 346.

RECENT RESULTS ON LATTICE GAUGE-HIGGS MODELS WITH FERMIONS

Robert E. Shrock

Institute for Theoretical Physics
State University of New York
Stony Brook, NY 11794 USA

ABSTRACT

Recent results on lattice gauge-Higgs models with fermions are discussed. Implications for electroweak theories are noted.

1. INTRODUCTION

Recently, we have shown that the (zero-temperature) chiral transition is an important feature of lattice gauge theories with Higgs and fermions fields.[1-10] Following initial numerical observations for SU(2) (with quenched fermions),[1] the existence, approximate location, and order of the chiral transition were determined analytically for both U(1) and Z(N)[2] and for SU(2)[6,7] using mean field methods applicable in the strong gauge coupling limit. The SU(2) theory with $I = 1/2$ Higgs fields and $I = 1/2$ fermions is of particular physical interest since it constitutes a lattice formulation of the SU(2) sector of the standard SU(2) $\times$ U(1) electroweak theory. Our analytic methods were shown[3] to go beyond the quenched approximation and include some effects of dynamical fermions. Although the analytic studies applied for $\beta_g = 0$, the behavior of the theory in a strip adjacent to the $\beta_g = 0$ edge of the phase diagram is qualitatively the same as that at this edge, owing to the finite radius of convergence of a small- β_g expansion; hence, the chiral transition point at $\beta_g = 0$ is the endpoint of a line of chiral transitions which extends into the interior of the diagram.[3] As was discussed in Ref. [3], this line cannot end anywhere in the interior, (since otherwise one could analytically continue the function representing $< \bar{\chi}\chi >$ from the chirally symmetric phase, where it vanishes identically, to the phase with spontaneous chiral symmetry breaking, where it is nonvanishing, which would be a contradiction); hence the chiral boundary completely separates the phase diagram into two (sets of) phases characterized by whether or not chiral symmetry is spontaneously broken. Since gauge-Higgs theories with Higgs fields transforming according to the fundamental representation have analytically connected confinement and Higgs phases, the chiral transition means that the addition of fermions to such theories significantly changes their phase structure.

The origin of the chiral transition can be explained as follows: spontaneous chiral symmetry breaking is due to the gauge-fermion interaction. In a theory with Higgs in the fundamental representation, as the gauge-Higgs coupling β_h increases, the gauge-Higgs interaction term has the effect of increasingly restricting the gauge field configurations which make a significant contribution to the path integral. This weakens the effective gauge-fermion interaction, and, as we showed both analytically and numerically, eventually the latter is rendered too weak to break the chiral symmetry spontaneously, and this symmetry is restored via the chiral phase transition.

2. IMPLICATIONS OF THE CHIRAL TRANSITION

The chiral transition has important implications for the electroweak theory:

(1) It is clearly important to have a formulation of electroweak interactions which treats the full path integral defining the quantum theory, and which subsumes, but is not limited to, perturbation theory. The lattice formulation provides a powerful means for making progress toward this goal. It is especially useful for investigating nonperturbative phenomena. The chiral transition is an example of this type of phenomenon, since the bare mass of the fermion is zero. It should be noted that the well-known problems associated with attempts to put chiral fermions on the lattice do not prevent one from using lattice techniques to study at least the SU(2) sector of the standard electroweak theory.[2,6] The reason for this is that one can write this theory in vectorlike form, owing to the fact that the representations of SU(2) are real. (However, since the representations of U(1) are complex, the full SU(2) × U(1) theory cannot be rewritten in such a manner.)

(2) A prerequisite for taking the continuum limit of the lattice theory is to determine the phase structure of the latter. Although gauge-Higgs theories with Higgs in the fundamental representation have analytically connected confinement and Higgs phases, this is no longer the case when fermions are included,

owing to the existence of the chiral transition.

(3) The continuum limit should be taken in the phase without spontaneous breaking of the chiral symmetry, since in the broken phase(s) the fermions pick up large, dynamically generated masses of order the electroweak scale of ~ 250 GeV. [2,4]

(4) It is thus of basic importance to determine where this phase without spontaneous chiral symmetry breaking lies in the SU(2) case with $I = 1/2$ Higgs and fermions. In addition to the analytic work which applies at $\beta_g = 0$ and, qualitatively, in the adjacent strip, the phase structure of this SU(2) theory has been mapped out by simulations using quenched[1,6,11] and dynamical[8] fermions.

(5) The line of chiral transitions which completely separates the chirally symmetric and spontaneously broken phases obviously alters the renormalization group flows in the theory. It is clearly important to understand the renormalization group flows and their fixed points with fermions included. This is necessary for a satisfactory understanding of the continuum limit of the lattice theory.

(6) Our work on the chiral transition may provide the first $ab\ initio$ answer[4] to the longstanding question of why the known weak nonsinglet fermions have $I = 1/2$. The point is that the strength of the effective gauge-fermion interaction increases with the I of the fermion. It was thus expected theoretically, and has been demonstrated analytically[2,6] and numerically[1,6] that the chiral boundary occurs at larger values of β_h for $I \geq 1$ fermions than for $I = 1/2$ fermions. It follows that if, as is now conventional, one takes an effective continuum limit from within the Higgs phase, in the vicinity of the O(4) transition at $\beta_g = \infty$, then this can be done while in the chirally symmetric phase for fermions with $I = 1/2$. However, since the chiral boundary for $I \geq 1$ fermions lies at larger β_h than that for $I = 1/2$ fermions, one would presumably be in the chirally broken phase for the former. Hence, only $I = 1/2$ fermions could be light compared to 250 GeV.

(7) In the case of SU(2), the lattice can be used to study the strongly coupled standard model.[2,13] This model requires that chiral symmetry not be broken spontaneously, since this would produce large, dynamically generated masses

for the fermions of order 250 GeV (and would violate electric charge); at the same time, however, it also assumes a strongly coupled, confined behavior with the resultant spectrum of excited bound states. It was motivated in part by early work with bosonic gauge-Higgs models with Higgs in the fundamental representation, which showed that the confinement and Higgs phases are analytically connected, giving rise to the principle of complementarity. However, because of the chiral transition separating the old confinement and Higgs phases, this complementarity does not hold in the presence of fermions. Moreover, in the phase with no spontaneous chiral symmetry breaking, numerical simulations have not been able to observe evidence of strongly coupled excited bound states.[1,11] (Parenthetically, one might note that in the full SU(2) × U(1) electroweak theory, even without fermions, there is a non-analytic boundary completely separating the interior of the phase diagram into two phases.[14]

3. BASIC TECHNIQUES AND RESULTS FOR U(1) AND SU(2)

In order to explain these results, we consider first a general lattice gauge theory with a Higgs field ϕ in the fundamental representation and a fermion field χ; this theory is described, in standard notation, by the path integral

$$Z = \int \prod_{n,\mu} dU_{n,\mu} d\phi_n d\phi_n^\dagger d\chi_n d\bar{\chi}_n e^{-S} \tag{1}$$

where $S = S_G + S_H + S_F$,

$$S_G = \beta_g \sum_{plaq.} [1 - P] \tag{2}$$

$$S_H = -2\beta_h \sum_{n,\mu} Re\{\phi_n^\dagger D_\phi(U_{n,\mu})\phi_{n+e_\mu}\} \tag{3}$$

$$S_F = \frac{1}{2} \sum_{n,\mu} \eta_{n,\mu} \bar{\chi}_n \Big[D_\chi(U_{n,\mu})\chi_{n+e_\mu} - D_\chi(U_{n-e_\mu,\mu}^\dagger)\chi_{n-e_\mu} \Big] \tag{4}$$

Here, $\beta_g = 1/g^2$ for U(1) and $4/g^2$ for SU(2); $P = Re\{U_{plaq.}\}$ for U(1) and $(1/2)Tr\{U_{plaq.}\}$ for SU(2). $D_\phi(U)$ is the representation of the gauge group element U corresponding to the Higgs field ϕ, and similarly for $D_\chi(U)$. The lattice is d-dimensional hypercubic with $d \geq 3$ and lattice spacing $a \equiv 1$. The Higgs fields satisfy $\phi_n^\dagger \phi_n = 1$; as is well known, this does not imply that the continuum Higgs fields are of fixed length. With no loss of generality, we take $\beta_h \geq 0$. We use staggered fermions χ_n because of their advantages for studies of chiral symmetry. The $\eta_{n,\mu} = \{1$ for $\mu = 1; \quad (-1)^{n_1 + \cdots + n_{\mu-1}}$ for $2 \leq \mu \leq 4\}$ are factors from the Dirac matrices. Although the fermions are coupled in a vectorlike manner to the gauge fields, this does not reduce the physical relevance of the model, since the SU(2) sector of the standard electroweak theory can be rewritten in a vectorlike manner. The U(1) theory is also useful, since it provides a simple laboratory in which to study chiral symmetry breaking without some of the algebraic complexities of SU(2) and exhibits results at strong gauge coupling which are similar to those for SU(2).

The analytic strong gauge-coupling results are obtained using the following technique. First, we perform the integrals over Higgs and gauge fields, obtaining the exact result

$$Z = J_1^{N_\ell} \int \prod_n d\chi_n d\bar{\chi}_n e^{-S'} \tag{5}$$

where N_ℓ is the number of links in the lattice, and $J_1 = I_0(2\beta_h)$ for U(1) and $I_1(2\beta_h)/\beta_h$ for SU(2), where $I_\nu(x)$ is the modified Bessel function. For U(1) we find[2]

$$-S'_{U(1)} = \sum_{n,\mu}[-(1/2)\eta_{n,\mu}r_q(\bar{\chi}_n\chi_{n'} - \bar{\chi}_{n'}\chi_n) + (1/4)(1 - r_q^2)\bar{\chi}_n\chi_n\bar{\chi}_{n'}\chi_{n'}] \tag{6}$$

where $n' = n + e_\mu$ and

$$r_q = I_q(y)/I_0(y) \tag{7}$$

where q denotes the fermion charge. r_q increases from 0 to 1 as β_h increases from 0 to ∞ and serves as an equivalent variable. For SU(2), S' has the more complicated form.[6] In the physically most interesting case of $I = 1/2$ fermions

$$\begin{aligned}
-S'_{SU(2),\, I=1/2} = \sum_{n,\mu}\Big[&-2^{-1}\eta_{n,\mu}r_{1/2}(\bar{\chi}_n\chi_{n'} - \bar{\chi}_{n'}\chi_n) \\
&+ 2^{-3}(1 - r_1)(\bar{\chi}_n\chi_n)(\bar{\chi}_{n'}\chi_{n'}) \\
&+ 2^{-2}(r_{1/2}^2 - r_1)(\bar{\chi}_n\chi_{n'})(\bar{\chi}_{n'}\chi_n) \\
&+ 2^{-3}(1 - r_{1/2}^2)\{(\bar{\chi}_n\chi_{n'})^2 + (\bar{\chi}_{n'}\chi_n)^2\} \\
&+ 2^{-4}\eta_{n,\mu}r_{1/2}a_6(\bar{\chi}_n\chi_{n'} - \bar{\chi}_{n'}\chi_n)(\bar{\chi}_n\chi_n)(\bar{\chi}_{n'}\chi_{n'}) \\
&+ 2^{-7}a_8(\bar{\chi}_n\chi_n)^2(\bar{\chi}_{n'}\chi_{n'})^2\Big]
\end{aligned} \tag{8}$$

where

$$r_I = I_{d_I}(y)/I_1(y) \tag{9}$$

$$a_6 = 1 - 4r_{1/2}^2 + 3r_1 \tag{10}$$

$$a_8 = -1 + 4r_{1/2}^2 - 3r_1^2 + 12r_{1/2}^2(r_1 - r_{1/2}^2) \tag{11}$$

Again, this result is exact. (Eq. (8) is written in an equivalent form in Ref. [6], using an identity given there.)

Next, we use a mean field technique to replace operator products higher than quadratic by a quadratic part times the mean value of the rest. This renders the action quadratic, so that we can perform the fermionic integrations exactly. We then derive a mean field consistency condition for the chiral symmetry breaking fermion condensate. In the U(1) case this equation is

$$< \bar{\chi}\chi > = (d/2)(1 - r_q^2) < \bar{\chi}\chi > \int (dp)D^{-1} \tag{12}$$

where $\int(dp) \equiv (2\pi)^{-d} \prod_{\mu=1}^{d} \int_{-\pi}^{\pi} dp_\mu$ and

$$D = [\{(d/2)(1 - r_q^2) < \bar{\chi}\chi >\}^2 + r_q^2 s] \tag{13}$$

with $s = \sum_{\mu=1}^{d} \sin^2 p_\mu$. As usual, (12) always has the solution $< \bar{\chi}\chi > = 0$, but if there is a solution for nonzero $< \bar{\chi}\chi >$, this minimizes the mean field free energy and hence is the physical solution. We have shown analytically that (12) implies the existence of a chiral transition in the theory.[5-7] The idea of the proof is to show first that for sufficiently large β_h, (12) has only the solution $< \bar{\chi}\chi > = 0$, while for $\beta_h = 0$ and in a neighborhood of this point, (12) has a nonzero solution. (The solution at $\beta_h = 0$ is $< \bar{\chi}\chi > = (2/d)^{1/2}$.) It then follows that $< \bar{\chi}\chi >$ must vanish non-analytically at a finite value of β_h and remain identically zero for larger values of β_h. Thus, the theory must undergo a phase transition at this point.

We have solved (12) numerically and have shown that the transition is continuous. A $1/d$ expansion of (12) to order $1/d^5$ has been calculated and gives excellent agreement with the numerical solution. This expansion was presented in Ref. [2] and was explicitly solved for the condensate in Ref. [9]. The solutions (analytic and numerical) yield a continuous transition. The $1/d$ expansion for the critical point is,[2,9] in terms of the variable $x \equiv r_q^2$,

$$x_c = \frac{1}{2}\left[1 + \frac{1}{4d} + \frac{5}{(4d)^2} + \frac{37}{(4d)^3} + \frac{365}{(4d)^4} + \frac{4373}{(4d)^5} + O(d^{-6})\right] \tag{14}$$

For $d = 4$, this gives the result $(r_q^2)_c = 0.55$, in excellent agreement with the numerical solution of (12). Indeed, in the large$-d$ limit, we have solved (12) exactly; since mean field theory is presumably exact in this limit, this should also consitute a solution of the full theory. In terms of the scaled condensate $\bar{M} = (d/2)^{1/2} M$, we obtain

$$\bar{M}^2 = \left[\frac{1 - 2x}{(1 - x)^2} \right] \theta(1/2 - x) \tag{15}$$

The $1/d$ expansion for $\bar{M}^2$ implies that the critical exponent for the condensate (i.e., the order parameter for the chiral transition) is $\beta = 1/2$. This value is the typical mean field prediction for spin systems; for such spin systems, this prediction yields the exact algebraic value of this exponent for $d \geq 4$.

For SU(2), the critical point can be located by an equation[6] similar to (14), with $x = r_{1/2}^2$ given by (9). This is a great simplification, since for fermions with higher I, the action S' becomes increasingly complicated, containing operator products up to order $4(2I + 1)$ in fermion fields.

A particularly important property of the mean field method used here is that it is not restricted to the quenched approximation. This can be seen as follows: given that the theory has a phase transition at $\beta_{h,c}$, the free energy is non-analytic there. It follows that all derivatives of the free energy are also non-analytic at the critical point. Since the average gauge-Higgs interaction $< L > = Re\{\phi_n^\dagger U_{n,\mu} \phi_{n+e_\mu}\}$ is given by $< L > = (2d)^{-1} \partial f / \partial \beta_h$ (where the reduced free energy is $f = \lim_{vol. \to \infty} N_s^{-1} \ln Z$), it follows that $< L >$ is non-analytic at $\beta_{h,c}$. In the quenched approximation, in contrast, there is no fermion feedback to the bosonic sector, so that $< L >$ would be the same function as in the pure bosonic gauge-Higgs theory without fermions (namely, $I_1(2\beta_h)/I_0(2\beta_h)$ for U(1) and $I_2(2\beta_h)/I_1(2\beta_h)$ for SU(2)) and hence would be analytic.

In both the U(1) and SU(2) cases, these analytic results obtained with approximate mean field methods have been compared with numerical results at $\beta_g = 0$, from both quenched fermions (Ref. [3] for U(1); Ref. [6] for SU(2)) and dynamical fermions (Ref. [12] for U(1); Ref. [10] for SU(2)). In all cases, the mean field approximation does at least as well as the typical $10 - 20\%$ accuracy typical of its use in $d = 4$ spin systems. The order of the chiral transition at $\beta_g = 0$ is found to be continuous, again in agreement with the mean field result. A detailed discussion of the simulation of the SU(2) lattice gauge-Higgs theory with $I = 1/2$ Higgs and

$I = 1/2$ dynamical fermions,[8] using the Langevin method, is given by I-H. Lee at this conference. Finally, one can get a rough determination of the critical exponent from the Monte Carlo data, and it is consistent with the mean field prediction $\beta = 1/2$.

As noted, these results should apply qualitatively in a strip adjacent to the $\beta_g = 0$ axis. This is found to be true from the quenched and dynamical Monte Carlo data. The chiral boundary has also been mapped out in the interior of the phase diagram for U(1)[3,10,12] and SU(2).[1,6,11] As expected from the analytic results at $\beta_g = 0$ and the physical reasons given above, it is found that for a given β_g, the critical value $\beta_{h,c}$ is an increasing function of q and I in the respective theories. It is also found that where the confinement-Higgs partial phase boundary is present, it coincides with the chiral boundary.

4. CHIRAL PHASE STRUCTURE FOR A THEORY WITH TWO DIFFERENT FERMIONS

Recently, we have extended these results in several ways.[9,10] First, we have used our analytic methods to determine the phase structure of a theory with two different types of interacting fermions simultaneously present. For technical simplicity, we used a U(1) theory and worked again at strong gauge coupling; however, we expect that, just as was true with our previous work, qualitatively similar results should hold in a strip adjacent to $\beta_g = 0$ and in the case of SU(2). Let us denote the different fermions as χ and ξ. Integrating over the Higgs and gauge fields as before, we obtain the exact result

$$Z = J_1^{N_l} \int \prod_n d\chi_n d\bar{\chi}_n d\xi_n d\bar{\xi}_n e^{-S'_{\chi\xi}} \tag{16}$$

$$\begin{aligned}
-S'_{\chi\xi} = \sum_{n,\mu} \Big[&-(1/2)\eta_{n,\mu} r_{q_\chi}(\bar{\chi}_n\chi_{n'} - \bar{\chi}_{n'}\chi_n) - (1/2)\eta_{n,\mu} r_{q_\xi}(\bar{\xi}_n\xi_{n'} - \bar{\xi}_{n'}\xi_n) \\
&+ 2^{-2}(1 - r_{q_\chi}^2)\bar{\chi}_n\chi_n\bar{\chi}_{n'}\chi_{n'} + 2^{-2}(1 - r_{q_\xi}^2)\bar{\xi}_n\xi_n\bar{\xi}_{n'}\xi_{n'} \\
&- 2^{-2}F_-(\bar{\chi}_n\chi_{n'}\bar{\xi}_{n'}\xi_n + \bar{\chi}_{n'}\chi_n\bar{\xi}_n\xi_{n'}) \\
&+ 2^{-2}F_+(\bar{\chi}_n\chi_{n'}\bar{\xi}_n\xi_{n'} + \bar{\chi}_{n'}\chi_n\bar{\xi}_{n'}\xi_n) \\
&+ 2^{-3}\eta_{n,\mu} r_{q_\chi} G\bar{\chi}_n\chi_n\bar{\chi}_{n'}\chi_{n'}(\bar{\xi}_n\xi_{n'} - \bar{\xi}_{n'}\xi_n) \\
&+ 2^{-3}\eta_{n,\mu} r_{q_\xi} G\bar{\xi}_n\xi_n\bar{\xi}_{n'}\xi_{n'}(\bar{\chi}_n\chi_{n'} - \bar{\chi}_{n'}\chi_n) \\
&- 2^{-4}a_8\bar{\chi}_n\chi_n\bar{\chi}_{n'}\chi_{n'}\bar{\xi}_n\xi_n\bar{\xi}_{n'}\xi_{n'} \Big]
\end{aligned} \tag{17}$$

where J_1 is defined for U(1) after (5) above, r_q is given by (7), and

$$F_+ = r_{q_\chi + q_\xi} - r_{q_\chi} r_{q_\xi} \tag{18}$$

$$F_- = r_{q_\chi - q_\xi} - r_{q_\chi} r_{q_\xi} \tag{19}$$

$$G = F_+ + F_- \tag{20}$$

$$a_8 = r_{q_\chi + q_\xi}^2 + r_{q_\chi - q_\xi}^2 - 4 r_{q_\chi} r_{q_\xi} (r_{q_\chi + q_\xi} + r_{q_\chi - q_\xi}) + 6 r_{q_\chi}^2 r_{q_\xi}^2 \tag{21}$$

Using mean field methods, we have shown that this theory exhibits two distinct (continuous) chiral transitions, and thus three different phases. The locations of the critical points for the two chiral transitions were also given. The phases are characterized as follows:

$$
\begin{aligned}
&P1: &\beta_h < \beta_{h,c,\xi}: &\quad <\bar\chi\chi> \neq 0; &\quad <\bar\xi\xi> \neq 0 \\
&P2: &\beta_{h,c,\xi} \leq \beta_h < \beta_{h,c,\chi}: &\quad <\bar\chi\chi> \neq 0; &\quad <\bar\xi\xi> = 0 \\
&P3: &\beta_h \geq \beta_{h,c,\chi}: &\quad <\bar\xi\xi> = 0; &\quad <\bar\chi\chi> = 0
\end{aligned}
\tag{22}
$$

The realization of chiral symmetry in these phases is as follows. Since $m_\chi = m_\xi = 0$, the original action is invariant under the continuous global symmetry $\prod_{f=\chi,\xi}$ $U(1)_{f,e} \times U(1)_{f,o}$, where "e" and "o" denote even and odd lattice sites, and each group $U(1)_{f,e} \times U(1)_{f,o}$ is generated by the global transformations

$$
\begin{aligned}
f_{n_e} &\to \exp(i\alpha_{f,e}) f_{n_e} \\
f_{n_o} &\to \exp(i\alpha_{f,o}) f_{n_o} \\
\bar f_{n_e} &\to \exp(-i\alpha_{f,o}) \bar f_{n_e} \\
\bar f_{n_o} &\to \exp(-i\alpha_{f,e}) \bar f_{n_o}
\end{aligned}
\tag{23}
$$

for $f = \chi, \xi$. This symmetry is realized explicitly in phase P3. It is spontaneously broken to $U(1)_{\chi,diag.} \times U(1)_{\xi,e} \times U(1)_{\xi,o}$ in phase P2, where the diagonal group $U(1)_{f,diag}$ is generated by the restriction of (23) to $\alpha_{f,e} = \alpha_{f,o}$. Here, the $U(1)_{\chi,diag}$ group expresses the global conservation of χ-type fermion number. Finally, in phase P1, the group $\prod_{f=\chi,\xi} U(1)_{f,e} \times U(1)_{f,o}$, is spontaneously broken to $U(1)_{\chi,diag} \times U(1)_{\xi,diag}$.

5. THEOREM FOR U(1) WITH ARBITRARY HIGGS AND FERMION CHARGES

Although in the standard electroweak theory, the Higgs field transforms according to the fundamental representation of SU(2), theories with higher unification require Higgs in higher representations. For example, in SU(5) grand unified theories, one needs an adjoint Higgs (24-dimensional representation), as well as a Higgs field in the fundamental (5) representation, and probably also a 45-dimensional Higgs. It is thus of interest to investigate the nature of chiral symmetry breaking in a model with a higher-dimensional Higgs representation. The U(1) gauge-Higgs-fermion theory provides a useful theoretical laboratory in which to investigate this question. Extending our original results in Ref. [2], we have now proved a general classification theorem which completely specifies under what conditions this theory has a chiral transition.[9] This result is exact and does not involve any mean field approximation. Although it applies at $\beta_g = 0$, it has implications also for the rest of the phase diagram. Consider the U(1) theory with a fermion f of charge q_f, and a Higgs field of charge q_h. Then

Theorem: If and only if q_f is a (nonzero) integral multiple of q_h, then the theory has a chiral phase transition.

If q_f is not a multiple of q_h, then the path integral factorizes into separate, decoupled bosonic and fermionic factors,

$$Z(q_f \neq kq_h) = Z_h Z_f \tag{24}$$

where k is a nonzero integer, $Z_h = I_0(2\beta_h)^{N_l}$, and

$$Z_f = \int \prod_n d\chi_n d\bar{\chi}_n \exp(2^{-2} \sum_{n,\mu} \bar{\chi}_n \chi_n \bar{\chi}_{n'} \chi_{n'}) \tag{25}$$

Since Z_f does not depend on β_h, it follows that $< \bar{\chi}\chi >$ is also independent of β_h; further, this condensate is given by its value at $\beta_h = 0$ in the pure gauge-fermion theory. Because the theory reduces to a $Z(N)$ gauge theory as $\beta_h \to \infty$, this leads to a corollary of the above theorem:

$$< \bar{\chi}\chi >_{U(1)} = < \bar{\chi}\chi >_{Z(N)} \qquad \text{at } \beta_g = 0 \tag{26}$$

Since the condensates in (26) are nonzero, it follows that if $q_f \neq kq_h$, chiral symmetry is spontaneously broken (at $\beta_g = 0$) for all β_h, including $\beta_h = \infty$.

6. YUKAWA INTERACTIONS IN U(1) AND SU(2) THEORIES

It is also clearly of interest to investigate the chiral transition in theories with Yukawa interactions. We have done this for U(1)[5] and recently for SU(2).[10] In both theories, the fermions consisted of a field χ in the fundamental representation interacting with gauge-singlet fermion(s) ξ (and ζ in the SU(2) case) via a Higgs in the fundamental representation. This is motivated by the Yukawa interactions in the SU(2) part of the standard electroweak theory. The analyses were carried out in the strong gauge coupling limit and used mean field techniques similar to those already discussed. In the U(1) case, the Yukawa interaction has the form

$$S_Y = h \sum_n (\bar{\chi}_n \phi_n \xi_n + h.c.) \tag{27}$$

In this model it was found that for a range of sufficiently weak Yukawa couplings h, the chiral transition continues to exist (and to be of second order), but is removed as the strength of the Yukawa coupling increases through a critical value, $h_c = (d/2)^{1/2}$. Interestingly, the neutral fermion, through its Yukawa interaction with the charged fermion, picks up a nonzero condensate $< \bar{\xi}\xi >$ in the broken phase. Both $\beta_{h,c}$ and $< \bar{\chi}\chi >$ decrease as h increases. One can observe an interesting balance in the sources of fermion masses: for zero Yukawa coupling, the physical fermions are massless (massive) in the respective phases where the condensates are zero (nonzero). As one turns on the Yukawa coupling, the region where the condensates vanish actually expands, but the fermions pick up masses directly from this coupling. Finally, when the strength of the Yukawa coupling exceeds the critical value, the fermion masses no longer get any dynamically generated contribution from the condensates, but instead arise entirely directly from the Yukawa interaction.

For the SU(2) case, there are two types of Yukawa interactions:

$$S_Y = \sum_n \left[h_1(\bar{\xi}_n \phi_n^\dagger \chi_n + h.c.) + h_2(\bar{\zeta}_n \epsilon(\chi_n \phi_n) + h.c.) \right] \tag{28}$$

where ϵ_{ij} is the totally symmetric tensor density. Using the same mean field theory methods as before (at $\beta_g = 0$), we find that if and only if

$$h_1^{-4} + h_2^{-4} > 8d^{-2} \tag{29}$$

the theory has a spontaneous chiral symmetry breaking phase transition. Note that the condition (29) is always satisfied if either h_1 or h_2 is smaller than $2^{-3/4}d^{1/2}$; in this case, the transition remains even for arbitrarily large values of the other

Yukawa coupling. As in the other cases, the transition was found to be continuous and approximate values for the critical point were given.

Acknowledgement

The research reported here was done in collaboration with I-H. Lee and, in Ref. [8], S. Aoki. This research was partially supported by the National Science Foundation under the grant NSF-PHY-8507627.

REFERENCES

1. I-H. Lee and J. Shigemitsu, Phys. Lett. **178B** (1986) 93.

2. I-H. Lee and R. E. Shrock, Phys. Rev. Lett. **59** (1987) 14.

3. I-H. Lee and R. E. Shrock, Nucl. Phys. **B290** (1987) 275.

4. I-H. Lee and R. E. Shrock, Phys. Lett. **196B** (1987) 82.

5. I-H. Lee and R. E. Shrock, Phys. Lett. **199B** (1987) 541.

6. I-H. Lee and R. E. Shrock, Phys. Lett. **201B** (1988) 497.

7. I-H. Lee, R. E. Shrock, in Proceedings of the International Symposium on Field Theory on the Lattice, Seillac, Nucl. Phys. B (Proc. Suppl.) 4 (1988), 373, 438.

8. S. Aoki, I-H. Lee, and R. E. Shrock, Phys. Lett. B, in press.

9. I-H. Lee and R. E. Shrock, BNL-Stony Brook preprint "Chiral Symmetry Properties of U(1) Lattice Gauge Theory with Scalar and Fermion Fields", BNL-ITP-SB-88-7 (April, 1988).

10. I-H. Lee and R. E. Shrock, BNL-Stony Brook preprint "The Chiral Transition in an SU(2) Gauge-Higgs-Fermion Theory with Yukawa Couplings", BNL-ITP-SB-88-27 (May, 1988).

11. A. De and J. Shigemitsu, Nucl. Phys. B, in press.

12. E. Dagotto and J. Kogut, Phys. Lett. B, in press.

13. M. Claudson, E. Farhi, and R. L. Jaffe, Phys. Rev. **D34** (1986) 873; L. Abbott and E. Farhi, Phys. Lett. **101B** (1981) 69.

14. R. E. Shrock, Phys. Lett. **B162** (1985); Nucl. Phys. **B267** (1986) 301; Phys. Rev. Lett. **56** (1986) 2124; Nucl. Phys. **B278** (1986) 380.

NON-PERTURBATIVE YUKAWA COUPLINGS

D. Stephenson and A. Thornton

Physics Department
University of Edinburgh
King's Buildings
Mayfield Road
Edinburgh EH9 3JZ
Scotland

ABSTRACT

We have calculated the phase diagram for a chiral Fermion-Higgs model on a 4^4 lattice using dynamical fermions. We find two phase transition lines for non-zero values of Yukawa coupling, which may prove useful in finding the continuum limit of the electroweak model.

1. INTRODUCTION

What is the phase diagram for a strongly Yukawa coupled Fermion-Higgs system? We have attempted to answer this question by performing a lattice simulation of a model containing dynamical fermions and radially-frozen Higgs fields. Previous studies[1-3] have used only quenched fermions which do not allow the fermions to act back upon the Higgs sector. This article is a summary of our work; more details are contained in [4].

2. THE MODEL

Hybrid Monte Carlo[5] was used to simulate the action for a radially-fixed U_1 model interacting with two species of naive fermions:

$$
S = - K \sum_{x,\mu} \left(\phi_x^+ \phi_{x+\hat{\mu}} + \phi_x^+ \phi_{x-\hat{\mu}} \right)
$$
$$
+ \sum_{x,y} \left(\overline{\psi}_x^{(1)} M_{xy} \psi_y^{(1)} + \overline{\psi}^{(2)} M_{xy}^+ \psi_y^{(2)} \right) .
\tag{1}
$$

The field $\phi_x = \exp(i\theta x)$ and the fermion matrix M_{xy} is given by:

$$M_{xy} = Y \exp\left(i\gamma_5\theta_x\right)\delta_{xy} + \frac{1}{2}\sum_\mu \gamma_\mu \left(\delta_{yx+\hat{\mu}} - \delta_{yx-\hat{\mu}}\right) . \tag{2}$$

The model has a global chiral U_1 symmetry which can be spontaneously broken by the Higgs sector. The masses of the fermions are generated by the non-zero Higgs field. Including only one species of fermion leads to a complex fermion determinant which cannot be easily studied by Monte Carlo techniques.

Two of the observables that were measured are:

$$
\begin{aligned}
P &= Tr(MM^+)^{-1}/N; \quad N = \text{number of lattice sites} \\
Q &= 0.5 + \sum_{x,\mu} \cos\left(\theta_{x+\hat{\mu}} - \theta_x\right)/8N .
\end{aligned}
\tag{3}
$$

A measure of the alignment of the Higgs field is given by Q which ranges from 0.5 for a disordered Coulomb phase up to 1.0 for an ordered Higgs phase. The fluctuations in Q correspond to a specific heat (CV) for the model and were also measured. A fermionic observable is given by the condensate P.

The hybrid Monte Carlo algorithm was implemented in OCCAM2 and ran on a processor—array made of 103 floating-point transputers. Identical copies of the algorithm were placed on 102 of the transputers and each were given different values of K and Y. On each transputer, 300 iterations were allowed for equilibration and then observables were measured over the next 1700 iterations. The algorithm was checked in a number of ways including comparing expectation values of observables with those expected from one-loop perturbative expressions calculated analytically.

3. THE RESULTS

The phase diagram that we obtained on a 4^4 lattice is shown in Figs. 1 and 2. Fig. 1 is a contour plot of the mean Q observable obtained from runs at 420 points evenly distributed in the area shown. Point A in the diagram is the 2nd order phase transition for the 4d XY model.[6,7] The fermions have the effect of inducing the bosons into a Higgs phase. For large Y, the fermions acquire large mass and so no longer affect the Higgs bosons. For values of K below that at A, this causes the appearance of two phase transition lines. The one going from A to B appears to be very strong but is difficult to investigate because it lies in an area of the diagram

254

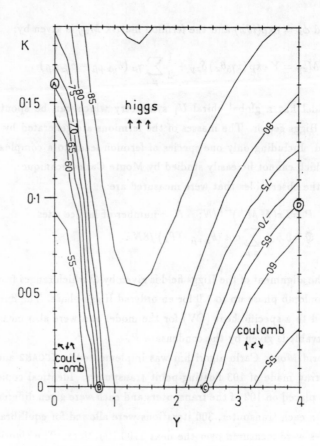

Figure 1. Dependence of $\langle Q \rangle$ on K and Y on a 4^4 lattice. Contours are labelled in units of 0.01 and are spaced in units of 0.05.

that is heavily infra-red affected. The one from C to D is described well by the $O(1/Y^2)$ expression:

$$K + Y^{-2} = K_c \text{ the value of K at A}. \tag{4}$$

Fig. 2 shows a contour plot for P. It appears strongly affected in the region of the first transition line and only slightly affected in the region of the second line. By estimating the position of peaks in the specific heat on the 4^4 lattice, we estimate A to be at $K = 0.15 \pm 0.1$, B at $Y = 0.64 \pm 0.1$ and C to be at $Y = 2.44 \pm 0.1$. To find these points more accurately and to determine the order of the transitions, we performed two runs on a 6^4 lattice. The results of the $K = 0$ run are shown in Fig. 3. On the 6^4 lattice, A was at $K = 0.16 \pm 0.1$, B was at $Y = 0.68 \pm 0.1$ and C was at $Y = 2.48 \pm 0.1$. These agree with those from the smaller lattice. By looking at the variation in the observables throughout the runs, no evidence was found for the existence of strongly 1st order transitions at points B and C.

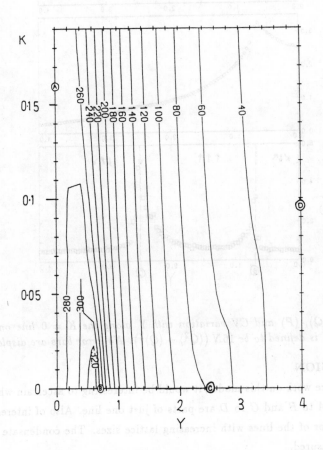

Figure 2. Dependence of $\langle P \rangle$ on K and Y on a 4^4 lattice. Contours are labelled in units of 0.01 and are spaced in units of 0.20.

256

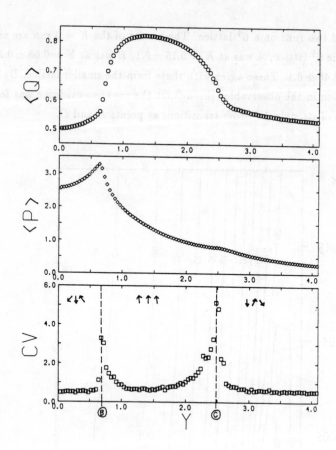

Figure 3. $\langle Q \rangle$, $\langle P \rangle$ *and CV variation with* Y *along the* $K = 0$ *line on a* 6^4
lattice. CV is defined to be $16N \left(\langle Q^2 \rangle - \langle Q \rangle^2 \right)$. *No error bars are displayed.*

4. DISCUSSION

In future work on this model, it would be interesting to ascertain whether or
not the lines A to B and C to D are parts of just one line. Also of interest would
be the behavior of the lines with increasing lattice sizes. The condensate TrM^{-1}
should be measured.

Extended models including Yukawa-Wilson terms, radially-free Higgs fields
and gauge fields could also be investigated.

257

Acknowledgements

We wish to thank Tien Kieu, Brian Pendleton and Jan Smit for useful discussions concerning this work.

This project required 300 hours on the Edinburgh Concurrent Supercomputer running at about 100 Mflops for which we acknowledge the support of SERC grant GR/E/21810, the Computer Board, the DTI and Meiko Ltd.

REFERENCES

1. J. Shigemitsu, Physics Letters 189B, (1987) 164.

2. J. Shigemitsu, Ohio preprint DOE/ER/01545-397.

3. A. De and J. Shigemitsu, Ohio preprint DOE/ER/01545-398.

4. D. Stephenson and A. Thornton, Edinburgh preprint 88/436.

5. S. Duane, A. D. Kennedy, B. J. Pendleton, and D. Roweth, Physics Letters 195B, (1987) 216.

6. K. C. Bowler, G. S. Pawley, B. J. Pendleton, D. J. Wallace, and G. W. Thomas, Physics Letters 104B, (1981) 481.

7. K. Jansen, J. Jersak, C. B. Lang, and G. Vones, Physics Letters 155B, (1985) 268.

MONTE CARLO COMPUTATION OF THE EFFECTIVE POTENTIAL FOR LATTICE ϕ^4 THEORY AND THE UPPER BOUND ON THE HIGGS MASS

M.M. Tsypin

Lebedev Physical Institute
Leninsky
Prospect 53
Moscow, USSR

In this report we describe a simple and efficient method of Monte Carlo investigation of scalar field theories. The idea is to compute the effective potential V_{eff} and to derive other properties of the theory (coupling constants, masses, etc.) from it. Instead of using external currents and Legendre transformation, we employ the definition of V_{eff} based on integration in the functional integral over all degrees of freedom except the only one corresponding to homogeneous field. It was suggested by J. Goldstone in the very first article[1] on the effective potential and is very convenient for Monte Carlo computations. With this technique we compute the upper bound on the renormalized coupling constant of the 4-component 0(4)-symmetric ϕ^4 theory in 4 dimensions in the broken phase and obtain the upper bound on the Higgs mass $m_H < 790 \div 850$ GeV.

Let us first consider the simplest possible scalar field theory: one-component ϕ^4 theory with Euclidean action

$$S = \int d^4x \mathcal{L} = \int d^4x \left\{ \frac{1}{2}\partial_\mu\phi\partial_\mu\phi + \lambda \left(\phi^2 - \eta^2 \right)^2 \right\} . \tag{1}$$

In the lattice regularized theory (see, e.g., [8]) ϕ is defined on the sites of the hypercubic lattic with spacing a, and

$$\int d^4x \rightarrow a^4 \sum_x, \quad \partial_\mu\phi \rightarrow \frac{1}{a} \left(\phi_{x+\hat{\mu}} - \phi_x \right) . \tag{2}$$

where $x + \hat{\mu}$ denotes the lattice site—the nearest neighbour of x in direction μ. The lattice action is

$$S = a^4 \sum_x \left\{ \frac{1}{2} \sum_\mu \left(\frac{\phi_{x+\hat\mu} - \phi_x}{a} \right)^2 + \lambda \left(\phi_x^2 \eta^2 \right)^2 \right\}, \tag{3}$$

where λ, η are the bare parameters. Expressing ϕ, η in units $\frac{1}{a}$, we get

$$S[\phi] = \sum_x \left\{ \frac{1}{2} \sum_\mu \left(\phi_{x+\hat\mu} - \phi_x \right)^2 + \lambda \left(\phi_x^2 - \eta^2 \right)^2 \right\}. \tag{4}$$

In the standard definition of the effective potential[2-7] we apply the external current $\mathcal{J}$ and compute the expectation value of ϕ:

$$\phi_{cl} \equiv <\phi>_{\mathcal{J}} = \frac{\int \prod_x d\phi \, \phi e^{-S[\phi] - \sum_x \mathcal{J}\phi_x}}{\int \prod_x d\phi \, e^{-S[\phi] - \sum_x \mathcal{J}\phi_x}}. \tag{5}$$

Then the Legendre transformation is performed to obtain V_{eff} obeying

$$\frac{dV_{\text{eff}} \left(\phi_{cl} \right)}{d\phi_{cl}} = -\mathcal{J}. \tag{6}$$

Alternatively, one may integrate this equation directly[5] to obtain V_{eff}. It is known,[5-7] that V_{eff} defined this way, must be convex, so the double-well shape is impossible. This is true for both infinite- and finite-volume systems.

To get rid of this unpleasant property, we use the different definition of V_{eff} .[1,9-11] Let us integrate in the functional integral over all degrees of freedom except the only one corresponding to momentum zero. This one degree of freedom we keep fixed and equal to $\hat\phi$:

$$\frac{1}{N} \sum_x \phi_x = \hat\phi, \tag{7}$$

where N is the number of lattice sites. The result depends on $\hat\phi$ only:

$$\int \prod d\phi \, e^{-S[\phi]} \bigg|_{\frac{1}{N} \sum_x \phi_x = \hat\phi} \equiv \int \prod d\phi \, \delta \left(\hat\phi - \frac{1}{N} \sum_x \phi_x \right) e^{-S[\phi]}$$

$$= e^{-S_{\text{eff}} \left(\hat\phi \right)} = e^{-\sum_x V_{\text{eff}} \left(\hat\phi \right)} = e^{-N V_{\text{eff}} \left(\hat\phi \right)}. \tag{8}$$

When we know V_{eff} defined this way, the vacuum expectation value of any operator X that depends on $\frac{1}{N} \sum_x \phi_x$ only can be easily computed:

$$\left\langle X\left(\frac{1}{N}\sum_x \phi_x\right)\right\rangle = \frac{\int \prod d\phi\, X\left(\frac{1}{N}\sum \phi_x\right) e^{-S[\phi]}}{\int \prod d\phi\, e^{-S[\phi]}}$$

$$= \frac{\int d\hat{\phi}\, X\left(\hat{\phi}\right) \int \prod d\phi\, \delta\left(\hat{\phi} - \frac{1}{N}\sum \phi_x\right) e^{-S[\phi]}}{\int d\hat{\phi} \int \prod d\phi\, \delta\left(\hat{\phi} - \frac{1}{N}\sum \phi_x\right) e^{-S[\phi]}} \tag{9}$$

$$= \frac{\int d\hat{\phi}\, X\left(\hat{\phi}\right) e^{-N V_{\text{eff}}\left(\hat{\phi}\right)}}{\int d\hat{\phi}\, e^{-N V_{\text{eff}}\left(\hat{\phi}\right)}}.$$

This formula gives us a simple method for Monte Carlo computation of V_{eff}. Consider the theory on a lattice with N sites. Let us generate the sequence of configurations by the usual Monte Carlo procedure. For each configuration compute $\hat{\phi} = \frac{1}{N}\sum_x \phi_x$. Thus we get the distribution of $\hat{\phi}$, described by the probability density $P\left(\hat{\phi}\right)$. From (9) we see that

$$P\left(\hat{\phi}\right) \sim e^{-N V_{\text{eff}}\left(\hat{\phi}\right)}, \tag{10}$$

and V_{eff} can be computed directly from the $\hat{\phi}$ distribution.

The example of such Monte Carlo computation is given at Fig. 1. It presents the $\hat{\phi}$ distribution for one-component theory. The action is

$$S[\phi] = \sum_x \left\{\frac{1}{2}\sum_\mu \left(\phi_{x+\hat{\mu}} - \phi_x\right)^2 + \lambda\left(\phi_x^2 - \eta^2\right)^2\right\}, \tag{11}$$

with bare parameters

$$\lambda = 10, \eta^2 = 0.3. \tag{12}$$

The $\hat{\phi}$ distribution of Fig. 1 is rather well approximated by

$$P\left(\phi\right) \sim e^{-N V_{\text{eff}}\left(\phi\right)}$$

$$V_{\text{eff}} = \lambda_{\text{eff}}\left(\phi^2 - \eta_{\text{eff}}^2\right)^2 \tag{13}$$

$$\lambda_{\text{eff}} = 1.4 \pm 0.15 \tag{14}$$

$$\eta_{\text{eff}} = \begin{cases} 0.165, & N = 6^4 \\ 0.155, & N = 8^4 \end{cases}.$$

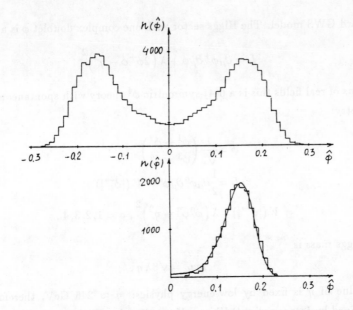

Figure 1. The $\hat{\phi}$ distribution for the one-component ϕ^4 theory ($\lambda =$ 10, $\eta^2 = 0.3$) on a 6^4 lattice (110000 configurations) and on an 8^4 lattice (16000 configurations). The curve corresponds to $\lambda_{eff} =$ 1.4, $\eta_{eff} = 0.155$.

The discrepancy in η_{eff} is a finite size effect. This system gives an example of great renormalization of parameters of the potential due to radiative corrections. Note that λ_{eff} is seven times smaller than the bare coupling λ, demonstrating the screening of bare charge in the ϕ^4 theory. It occurs in this system with spontaneously broken symmetry just the same way as in the symmetric case.[12]

It is interesting that although the coupling constant is large, the effective potential is rather well approximated by the classical formula (13) (with parameters entirely different from the bare ones), and there is no visible distortion of its shape by radiative corrections.

The methods described above can be easily applied to the Higgs sector of the standard GWS model. The Higgs sector with one complex doublet ϕ is a ϕ^4 theory:

$$L = \partial_\mu \phi^+ \partial_\mu \phi + \lambda \left(2\phi^+ \phi - \eta^2 \right)^2 . \tag{15}$$

In terms of real fields this is a $0(4)$-symmetric ϕ^4 theory with spontaneously broken symmetry:

$$\phi = \frac{1}{\sqrt{2}} \begin{pmatrix} \phi^1 + \phi^2 \\ \phi^3 + \phi^4 \end{pmatrix} \tag{16}$$

$$L = \frac{1}{2} \partial_\mu \phi^a \partial_\mu \phi^a + V \left(|\vec{\phi}| \right) \tag{17}$$

$$V \left(|\vec{\phi}| \right) = \lambda \left(\phi^a \phi^a - \eta^2 \right)^2 , \ a = 1, 2, 3, 4 . \tag{18}$$

The Higgs mass is

$$m_H = \sqrt{8\lambda}\, \eta . \tag{19}$$

The value of η is fixed by low-energy physics: $\eta \approx 246$ GeV, therefore m_H is determined by (renormalized) Higgs self-coupling λ_R. The upper bound on λ_R will set the upper bound on m_H:

$$m_H < \sqrt{8\lambda_{max}} \cdot 246 GeV . \tag{20}$$

The lattice computation of this upper bound was proposed in [13]. We perform it as follows.

The theory (17, 18), regularized by the lattice, has the action

$$S = \sum_x \left\{ \frac{1}{2} \sum_\mu \sum_a \left(\phi^a_{x+\hat{\mu}} - \phi^a_x \right)^2 + \lambda \left(\phi^a_x \phi^a_x - \eta^2 \right)^2 \right\} \tag{21}$$

(ϕ, η are expressed in units $1/a$). We set the bare coupling $\lambda = \infty$ and evaluate λ_R as a function of m_H. With $\lambda = \infty$ the action becomes that of a nonlinear sigma model:

$$S = -\sum_x \sum_\mu \phi^a_{x+\hat{\mu}} \phi^a_x, \qquad \phi^a \phi^a = \eta^2 . \tag{22}$$

For this model we generate the sequence of configurations by the Monte Carlo algorithm. To update ϕ^a in the lattice site the test value of ϕ^a distributed uniformly on a sphere $\phi^a \phi^a = \eta^2$ is generated. It is then accepted or rejected according to

Metropolis algorithm. We repeat this procedure twice at each lattice site. The new configuration is obtained after one sweep through the entire lattice. To obtain four-component vectors ϕ^a distributed uniformly on a sphere we generate points distributed uniformly inside the unit sphere and project them onto $\phi^a \phi^a = \eta^2$.

For every configuration we compute

$$\hat{\phi}^a = \frac{1}{N} \sum_x \phi_x^a \,.$$

Then the distribution of $\hat{\phi} = \sqrt{\hat{\phi}^a \hat{\phi}^a}$ is drawn. The probability density for 4-component vector $\hat{\phi}^a$ is proportional to

$$e^{-N V_{\text{eff}} \left(\hat{\phi} \right)} \,;$$

therefore $\hat{\phi}$ is distributed with density

$$P \left(\hat{\phi} \right) \sim \hat{\phi}^3 e^{-N V_{\text{eff}} \left(\hat{\phi} \right)} \,. \tag{23}$$

Parameters of the effective potential

$$V_{\text{eff}} \left(\hat{\phi} \right) = \lambda_{\text{eff}} \left(\hat{\phi}^2 - \eta_{\text{eff}}^2 \right)^2 \tag{24}$$

can be found by comparing (23) with Monte Carlo data. Examples are given in Figs. 2 and 3. To compare the Higgs mass m_H and the renormalized coupling λ_R one must take into account the renormalization of the kinetic part of the Lagrangian as well as of the static one. It leads to renormalization of the propagator and the field ϕ. To compute this renormalization one should retain in the effective action not only one degree of freedom corresponding to zero momentum, but also the degrees of freedom corresponding to small but nonzero momenta. This computation shows that the field renormalization constant is close to unity, and m_H and λ_R can be derived directly from V_{eff} :

$$\lambda_R \cong \lambda_{\text{eff}}, \qquad m_H \cong \sqrt{8 \lambda_{\text{eff}}} \, \eta_{\text{eff}} \,. \tag{25}$$

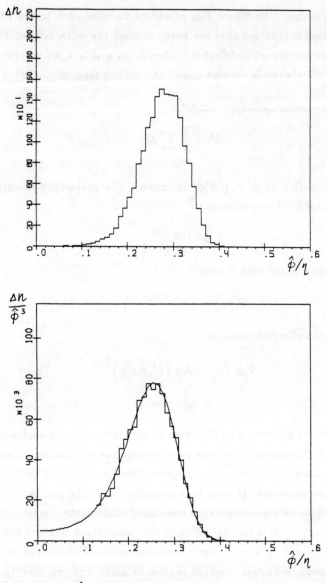

Figure 2. The $\hat{\phi}$ distribution for the 4-component sigma model on a 6^4 lattice for $\eta^2 = 0.62$ (18000 configurations). The curve corresponds to (23,24), $\lambda_{\text{eff}} = 1.3$, $\lambda_{\text{eff}} = 0.257\eta$.

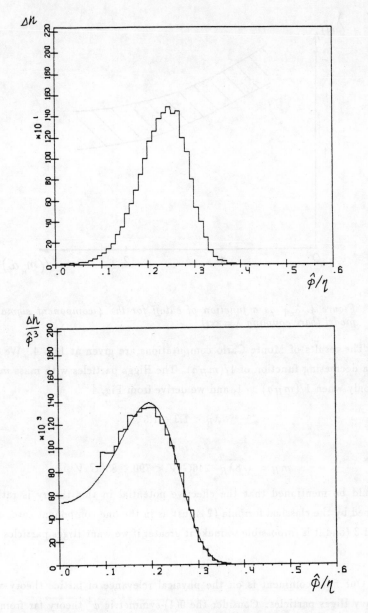

Figure 3. The $\hat{\phi}$ distribution for the 4-component sigma model on a 6^4 lattice for $\eta^2 = 0.61$ (19000 configurations). The curve corresponds to (23,24), $\lambda_{\text{eff}} = 1.3$, $\lambda_{\text{eff}} = 0.196\eta$.

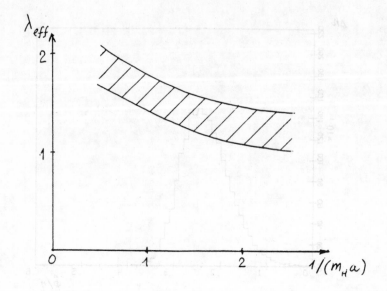

Figure 4. λ_{eff} as a function of cutoff for the 4-component sigma model (bare coupling $\lambda = \infty$).

The results of Monte Carlo computations are given at Fig. 4. We see that λ_R is a decreasing function of $1/(m_H a)$. The Higgs particles with mass m_H make sense only when $1/(m_H a) \gg 1$, and we derive from Fig. 4

$$\lambda_R < 1.3 \div 1.5 , \tag{26}$$

and

$$m_H = \sqrt{8\lambda_R} \cdot 246 GeV < 790 \div 850 GeV . \tag{27}$$

It should be mentioned that the effective potential in this theory is rather well described by the classical formula (24) just as in the one-component case, although $\lambda_R \approx 1.3$ (and it is impossible to make it greater if we want Higgs particles to make sense).

Our final comment is on the physical relevance of lattice theory with superheavy Higgs particles. Consider the $0(4)$-symmetric ϕ^4 theory far from critical point (in spontaneously broken phase). Then λ_R and m_H can be arbitrarily high, but $m_H \gg 1/a$. Lattice theories far from critical point are usually considered irrelevant for continuum physics. But in this theory there are 3 massless Goldstone particles, and it can be regarded as effective quantum field theory for these particles, applicable for momenta $p \ll 1/a$. Obviously, the superheavy Higgs particle cannot be created in their collisions. But m_H appears in V_{eff} (ϕ) and can be seen in corrections to low-energy Goldstone-Goldstone scattering, and this m_H is superheavy.

So our upper bound on m_H is for the Higgs particle that can be created as a distinct resonance, and not for the Higgs particle observed indirectly in low-energy scattering.

Acknowledgements

I would like to express my gratitude to M.B. Voloshin for calling my attention to the problem and valuable discussion and to V.Ya. Fainberg for valuable discussions.

REFERENCES

1. J. Goldstone, Nuovo Cim. 19, (1961) 154.

2. G. Jona-Lasinio, Nuovo Cim. 34, (1964) 1790.

3. S. Coleman, E. Weinberg, Phys. Rev. D7, (1973) 1888.

4. S. Coleman, R. Jackiw, H.D. Politzer, Phys. Rev. D10, (1974) 2491.

5. D.J.E. Callaway, D.J. Maloof, Phys. Rev. D27, (1983) 406.

6. D.J.E. Callaway, Phys. Rev. D27, (1983) 2974.

7. Y. Fujimoto, L. O'Raifeartaigh, G. Parravicini, Nucl. Phys. B212, (1983) 268.

8. I.G. Halliday, Rep. Progr. Phys. 47, (1984) 987.

9. P. Ginsparg, H. Levine, Phys. Lett. B131, (1983) 127.

10. M.M. Tsypin, Preprint 280 of P. N. Lebedev Physical Institute.

11. L. O'Raifeartaigh, A. Wipf, H. Yoneyama, Nucl. Phys. B271, (1986) 653.

12. B. Freedman, P. Smolensky, D. Weingarten, Phys. Lett. B113, (1982) 481.

13. R. Dashen, H. Neuberger, Phys. Rev. Lett 50, (1983) 1897.

SCALING LAWS AND TRIVIALITY BOUNDS IN THE LATTICE n-COMPONENT ϕ^4 THEORY

Peter Weisz

Supercomputer Computations Research Institute
Florida State University
Tallahassee, FL 32306-4052

ABSTRACT

A report is given on the status of a program to 'solve' the lattice n-component ϕ_4^4 theory, in a large region of bare parameter space, including, in particular, a region of weak coupling in the broken symmetry phase.

1. INTRODUCTION

This talk is essentially a progress report on the recent work that Martin Lüscher and I have been doing on the n-component ϕ_4^4 theory. Our initial motivation to begin our detailed analytical studies was to provide accurate estimates of various observables in the infinite volume theory which could be used as a control in a numerical study of finite volume effects, at first in the symmetric phase. It soon became apparent however that we could extend the scope of our investigation to solve the model in a large region of the bare parameter space including the entire symmetric phase and a region of the broken phase around the critical line. By 'solve' is meant the determination of low energy quantities to a (good) estimable accuracy. In particular we could carry out analytically the proposal of Dashen and Neuberger[1] to determine the maximum renormalized coupling (as a function of the bare coupling) for a fixed renormalized mass. There have been many Monte-Carlo studies of the latter program[2] and it will be interesting to compare our results with these, particularly since there are subtleties in determining (resonance) masses in (necessarily finite volume) numerical simulations, in the case that Goldstone bosons are present, which have so far been treated rather cavalierly.

The complete details and results of our calculations in the 1-component model have already been published.[3,4] We are now on the brink of completing the analysis of the physically more interesting case of $n = 4$.[5] Actually much of the calculation has been done for arbitrary n. We have also studied the $n \to \infty$ limit and found that although the same qualitative picture holds there, extrapolations of these results to small n values are often quantitatively unreliable. In these proceedings I will therefore be brief concerning the technical details and merely repeat the main stategy of our approach. I refer the interested reader to the original papers cited above and to the recent nice review by Martin Lüscher[6] in which the finite volume studies alluded to above are also discussed.

2. NOTATION AND NORMALIZATION CONDITIONS

We consider the theory of an n-component scalar field $\phi_\alpha(x), (\alpha = 1, ..., n)$, on a hypercubic lattice of infinite extent in all directions, specified by the action

$$S_h = S + h \sum_x \phi_n(x), \tag{1}$$

where,[1]

$$S_h = \sum_x \left\{ -\kappa \sum_{\mu=0}^3 (\phi(x) \cdot \phi(x + \hat{\mu}) + \phi(x) \cdot \phi(x - \hat{\mu})) \right. $$
$$\left. + \phi(x) \cdot \phi(x) + \lambda(\phi(x) \cdot \phi(x) - 1)^2 \right\}, \tag{2}$$

in the limit that the external field $h \to 0$. The bare parameters κ (the hopping parameter) and λ (the bare coupling are restricted to non-negative values. In the limit $\lambda \to \infty$ the model reduces to the non-linear sigma model and to the Ising model in the special case $n = 1$. Expectation values are defined in the conventional way.

In the symmetric phase we can (and it is convenient to) impose renormalization conditions on the vertex functions at zero external momenta in order to define a renormalized mass m_R and renormalized coupling g_R.[3] In the broken phase one cannot impose all conditions at zero momentum, and here our

---

[1] Here, and below, the lattice spacing a is equal to 1. $\hat{\mu}$ is the vector of a lattice link in the μ-direction.

definitions are as follows. The wave function renormalization constant Z_R is defined through the behaviour of the inverse propagator of the Goldstone bosons at zero momentum:

$$\Gamma^{(2)}(p, -p)_{ab} = -\delta_{ab} Z_R^{-1} \left\{ p^2 + O(p^4) \right\}, \qquad (p \to 0), \qquad (a, b < n). \qquad (3)$$

The vacuum expectation value v_R of the renormalized σ-meson field is then given by

$$v_R = v Z_R^{-1/2} . \qquad (4)$$

A renormalized mass m_R that we found convenient for our calculations is specified by a zero of the real part of the inverse σ-meson propagator,

$$\Re \left\{ \Gamma^{(2)}(p, -p)_{nn} \big|_{p=(im_R,0,0,0)} \right\} = 0 . \qquad (5)$$

Finally the definition of the renormalized coupling that we adopt is

$$g_R = 3 m_R^2 / v_R^2 . \qquad (6)$$

With this choice, the formula for the ratio of the Higgs to W-meson mass m_W when the O(4)-model is embedded in the full lattice Higgs model

$$\frac{m_R^2}{m_W^2} = \frac{4 g_R}{3 g^2} + O(g^0) , \qquad (7)$$

where g is the renormalized gauge coupling, is valid to all orders in g_R.[1,6]

3. PHYSICAL MASS AND UNITARITY BOUNDS

The physical σ-meson mass m_σ and width Γ_σ are defined through the position of the pole of the σ-meson propagator

$$\Gamma^{(2)}(p, -p)_{nn} \big|_{p=(im_\sigma + \frac{1}{2}\Gamma_\sigma, 0, 0, 0)} = 0 . \qquad (8)$$

For weak coupling, the renormalized mass m_R and the physical mass m_σ are numerically very close; to 2-loop renormalized perturbation theory they are related by,

$$m_\sigma = m_R \left\{ 1 + \frac{\pi^2}{288}(n-1)^2 \alpha_R^2 + O(g_R^3) \right\} , \qquad (9)$$

where

$$\alpha_R = \frac{g_R}{16\pi^2} \, . \tag{10}$$

The S-wave 'isospin' 0-channel partial wave amplitude of elastic scattering of the Goldstone bosons is to tree level given by,

$$t_0^0 = \frac{g_R}{48\pi\sqrt{s}} \left\{ (n-1)\frac{s}{m_R^2 - s} - 2 + 2\frac{m_R^2}{s} \ln\left(1 + \frac{s}{m_R^2}\right) \right\} \, , \tag{11}$$

where s is the usual Mandelstam variable.[2] Using the unitarity requirement

$$|\Re t_0^0| < \frac{1}{\sqrt{s}} \, , \tag{12}$$

we obtain the tree level unitarity bound

$$g_R < \frac{48\pi}{n+1} \, . \tag{13}$$

For the case $n = 4$ this implies $g_R < 30$, or using the phenomenological value $v_R = 250$ GeV one would get the bound $m_R < 800$ GeV. This is about half the value quoted by Lee, Quigg and Thacker[7]—a factor of 2 arising from the fact that these authors only impose the unitarity restriction $|t_0^0| < \frac{2}{\sqrt{s}}$, instead of the stronger requirement eq. (12).[3]

4. ANALYSIS STRATEGY

Our strategy of analysis of the lattice model follows in three steps. The first two steps deal with the solution of the model in the symmetric phase and use well-known established techniques. The third step, extending the analysis to the broken phase, contains the essential new ingredient. Here we only summarize the most important points.

Step 1: Deep in the symmetric phase quantities of interest are calculated to high orders in the "high-temperature" (i.e. small κ) expansion. For our purposes it

---

[2] The representation eq. (11) is only valid outside the resonance region i.e. for $|s - m_R^2| \gg m_R\Gamma_\sigma$.

[3] An extra factor $\frac{5}{6}$ arises from the fact that in Ref. [7] also 4-particle production is included in the unitarity balance.

suffices to calculate the susceptiblities χ_2 and χ_4 and the second-moment μ_2 defined by

$$\chi_2 \delta_{\alpha\beta} = \sum_x \langle \phi_\alpha(x)\phi_\beta(0) \rangle, \tag{14}$$

$$\chi_4(\delta_{\alpha\beta}\delta_{\gamma\delta} + \delta_{\alpha\gamma}\delta_{\beta\delta} + \delta_{\alpha\delta}\delta_{\beta\gamma})/3 = \sum_{x,y,z} \langle \phi_\alpha(x)\phi_\beta(y)\phi_\gamma(z)\phi_\delta(0) \rangle^{\text{conn}}, \tag{15}$$

$$\mu_2 \delta_{\alpha\beta} = \sum_x x^2 \langle \phi_\alpha(x)\phi_\beta(0) \rangle. \tag{16}$$

For the case $n = 1$ we were fortunate to be able to use the 10'th order series for these quantities published by Baker and Kincaid.[8] However, for the cases $n > 1$ no comparative long series have (to our knowledge) been published. Using the linked-cluster expansion we have thus calculated the series to 14'th order for general n.[9] [4] The results are available as computer files, and these can be readily obtained per BITNET from the authors on request.

λ	κ_c(our)	κ_c(their)	[ref.]
1.0	0.2468(1)	0.2471(9)	[11]
3.202	0.2831(1)	0.2829(6)	[14]
∞	0.3041(1)	0.3039(8)	[11]
∞	0.3041(1)	0.3038(2)	[12]

Table 1. Comparison between results for κ_c for the $n = 4$ model obtained in this analysis and various Monte-Carlo simulations.

The series are analysed and in particular serious effort is made to estimate the systematic errors. The critical line $\kappa = \kappa_c(\lambda)$ can be determined to a good accuracy by an analysis of the series for χ_2 and incorporating the singularity behaviour predicted by the renormalization group. A comparison of our results with some results obtained in numerical simulations[11,12,13] is shown in table 1. The series for g_R, m_R, and Z_R are extrapolated (for a fixed λ) to a κ-value where the renormalized coupling is $< \frac{2}{3}$ of the tree-level unitarity bound. In the symmetric phase this corresponds to a correlation length $\approx 2, 3$ for $n = 1, 4$ respectively, and thus, no

---

[4] for dimensions 2, 3 as well as 4

'wild' extrapolation to huge correlation lengths is involved, and estimated systematic errors are under control and reasonably small.

Step 2: In the remaining region of the symmetric phase where $g_R < \frac{2}{3}$ of the tree-level unitarity bound we assume the validity of renormalized perturbation theory. In these regions the $O(a^2)$ cutoff dependence of the perturbative coefficients is typically also relatively weak and hence the origin of reference to this region as the "scaling region." The values of g_R, m_R and m_R in the scaling region are then obtained by integration of the renormalization group equations describing their development along lines of fixed bare coupling λ, using as initial data the values at the boundary of the scaling region obtained in Step 1. In these equations we use renormalization functions calculated in our scheme to 3-loops and include the full cutoff dependence to 1-loop.

There are various consistency checks that can be made to test the validity of the hypotheses made in our analysis described above. Firstly one can calculate various low-energy physical quantities in perturbation series and compare the contributions at successive orders. We have done this for a variety of quantities and have generally 'observed covergence' of the series in the scaling region. A second and more convincing check is to compare results obtained in the neighborhood of the boundary of the scaling region by i) further extrapolating the high temperature expansions into the scaling region, with those by ii) integrating the renormalization group equations out of the scaling region. As an example, the results of such a comparison is shown in table 2 for the case $n = 4$ for $\lambda = \infty$; the matching is truly impressive.[5] A final check is to compare the results with accurate Monte-Carlo data at reasonably large correlation lengths. For the case $n = 4$ most of the simulations to date in this phase have been done for values of the bare parameters where the high-temperature expansion alone give a very good description of the data. In table 3 we compare some of the results of the simulation ref. [14] with ours—the agreement is satisfactory. For $n = 1$ the data from a recent precision numerical simulation of the Ising model[10] is also in concord with our expectations as illustrated in fig. 1. The error bars from our previous work[3] seem embarassingly large on this plot but firstly the reader should appreciate the scale of the figure and secondly that, with the

---

[5] equally good results are obtained for all λ values

longer high temperature series now at hand the analysis for $n = 1$ could be repeated and the error bars reduced. If the data of ref. [10] is accepted, then it appears that the true beta function lies somewhere between the 2- and 3-loop functions.

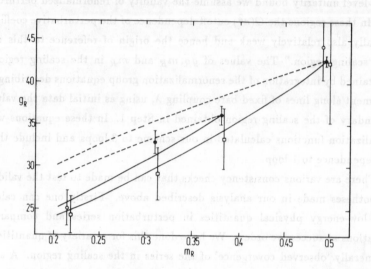

Figure 1. The renormalized coupling g_R as a function of m_R in the 4-dimensional Ising model in the symmetric phase. The solid circles are data of ref. [10] and the solid and dashed curves are results of integration of the renormalization group equations using 3- and 2- loop functions respectively starting at these data points. The figure is taken from ref. [10]—only our predictions ref. [3] at representative points (squares) and older data ref. [17] (open circles) have been superimposed.

In our analysis, the renormalized coupling g_R scales to zero as one approaches the critical line $\kappa = \kappa_c(\lambda)$ in the symmetric phase, in such a way that the limit

$$C_1(\lambda) = \lim_{\kappa \to \kappa_c} a m_R (\beta_1 g_R)^{\beta_2/\beta_1^2} e^{1/\beta_1 g_R} \qquad (17)$$

exists. This and related constants, which embody the remnants of non-perturbive information, are obtained with small estimable errors.

Step 3: In the broken phase there is no known practical low-temperature expansion which would give us accurate numerical information in a region where the correlation is small. However, in the neighborhood of the critical line there is, also in this phase, a scaling region where renormalized perturbation theory yields

λ	κ	$Volume\ V$	$m_R(their)(V)$	$m_R(our)(\infty)$
2.714	0.260487	8^4	0.61	0.600(2)
2.940	0.271107	8^4	0.43	0.410(2)
3.018	0.274675	8^4	0.36	0.342(2)
3.018	0.274675	12^4	0.33	0.342(2)
3.097	0.278256	12^4	0.27	0.252(2)

Table 2.   Comparison between results for m_R for the $n = 4$ model obtained in this analysis and data read off from the figures in ref. [14] (without estimated error).

a good approximation. The solution of the model in this domain relies on the observation that the theory in the scaling regions on both sides of the critical line can be related, by mass perturbation theory, to the critical theory, (as explained in detail in [4]). The scaling properties in the two phases can thereby be related to each other. In particular one finds that the constant $C_1'(\lambda)$ defined in the broken phase, in an analogous fashion to $C_1(\lambda)$ in eq. (17), is linearly related to the latter, the precise relation requiring only a 1-loop calculation.[4,5] Moreover, the renormalization constants approach the same values from either side of the critical line. With this information we integrate the renormalization group equations away from the critical line into the broken phase, treating the equations as we did in Step 2, until g_R reaches $\frac{2}{3}$ of the tree-level unitarity bound eq. (13).

5. CONCLUSION AND GENERAL REMARKS

The oft conjectured property of 'triviality' of the (lattice) ϕ^4 theory in 4-dimensions; i.e., the vanishing of the renormalized coupling as one approaches the critical line, is obtained in our treatment. Our analysis, of course, does not constitute a rigorous proof of this property but it does provide rather overwhelming evidence.

As already mentioned we are in the process of completing Step 3 above, however we do not expect any qualitative differences fom the case $n = 1$. In particular we expect that the minimum value of $m_R a$ along the renormalization group trajectories is attained in the limit $\lambda \to \infty$. Thus bounds on the ratio m_R/v_R for ranges of m_R in the scaling region are obtained by studying this limit. These

m_R	g_R	Z_R	κ
0.50	26.0(6)	1.717(4)	0.28705(8)
0.50	27(2)	1.705(9)	0.2870(2)
0.40	22.8(8)	1.682(5)	0.29247(8)
0.40	23(2)	1.676(8)	0.2925(1)
0.30	20(1)	1.653(6)	0.29708(7)
0.30	20(1)	1.652(7)	0.29708(7)
0.20	16(1)	1.631(7)	0.30071(5)
0.20	16.4(9)	1.634(7)	0.30072(9)
0.10	12(2)	1.616(8)	0.30315(2)
0.10	12.9(6)	1.622(7)	0.3031(1)

Table 3. Comparison between results for $\lambda = \infty, n = 4$, obtained from the high-temperature analysis (first row for a given m_R) and from integration of the renormalization group equations (second row), with initial values equated to the high-temperature data in the symmetric phase at $\kappa = 0.98\kappa_c$ at which $m_R \approx 0.3$.

triviality bounds are inherently non-universal, they depend on the regularization scheme. However it is hard to imagine that these bounds are hyper-sensitive to moderate changes of the regularization scheme.[6] The recent computations of Bhanot and Bitar[15] seem to confirm this picture. In connection with such calculations arises a question of principle: how can cutoffs of different regularization schemes be compared? One possible way would be to match $O(a^2)$ in a given processes for some value of the renormalized coupling in the scaling region; this would however only make sense if the scaling violations of one regularization were systematically larger

---

[6] I find the point of view on this situation expressed by Adler in his summary of this conference very reasonable.

than the other for a wide class of processes. Finally we also anticipate (at least for the regularized model studied) that again as in the $n = 1$ component model there will be no strongly interacting effective continuum theory. How 'universal' such a property is is also not known.

It will be interesting to compare our results with the published Monte-Carlo data.[12,13] However in the analysis of the numerical data, various definitions of σ-masses are made, some even ill-defined in the infinite-volume limit, and it is not a priori clear how these compare with our definition eq. (5). For the Monte-Carlo simulations of models with Goldstone particles there is still much to be learned concerning how to control of the systematic finite volume effects and in particular how to properly identify resonance parameters.[16] Our analytical results on the infinite volume limit should serve as a reliable monitor in such an endeavour.

Acknowledgements

I would like to thank Martin Lüscher for a most enjoyable collaboration and for sharing his deep physical and analytical insights with me.

REFERENCES

1. R. Dashen and H. Neuberger, Phys. Rev. Lett., 50, (1983) 1897.

2. J. Kuti, C.Lang and K. Bitar, See contributions in these proceedings and the references therein.

3. M. Lüscher and P. Weisz, Nucl. Phys., B290 [FS20], (1987) 25.

4. M. Lüscher and P. Weisz, Nucl. Phys., B295 [FS21], (1988) 65.

5. M. Lüscher and P. Weisz, in preparation.

6. M. Lüscher, "Solution of the lattice ϕ^4 theory in 4 dimensions," preprint DESY 87-159, Cargese Summer School Lectures 1987, to be published in the proceedings.

7. B. W. Lee, C. Quigg and H. B. Thacker, Phys. Rev., D16, (1977) 1519.

8. G. A. Baker and J. M. Kincaid, J. Stat. Phys., 24, (1981) 469.

9. M. Lüscher and P. Weisz, "Application of the linked cluster expansion to the n-component ϕ^4 theory," preprint FSU-SCRI-88-07, to be published in Nucl. Phys. B.

10. I. Montvay, G. Münster and U. Wolff, "Percolation cluster algorithm and scaling behaviour in the 4-dimensional Ising model," preprint DESY 88-049, to be published in Nucl. Phys. B.

278

11. W. Bernreuther, M. Göckeler and M. Kremer, Nucl. Phys., B295 [FS21], (1988) 211.

12. A. Hasenfratz, K. Jansen, C. B. Lang, T.Neuhaus and H. Yoneyama, Phys. Lett., 199B, (1987) 531.

13. J. Kuti, L. Lin and Y. Shen, "Upper bound on the Higgs mass in the standard model," preprint UCSD/PTH 87-18.

14. J. Kuti and Y. Shen, Phys. Rev. Lett., 60, (1988) 85.

15. G. Bhanot and K. Bitar, "Regularization dependence of the lattice Higgs mass bound," preprint FSU-SCRI-88-50.

16. U. Wiese, in preparation.

17. I. Montvay and P. Weisz, Nucl. Phys., B290 [FS20], (1987) 327.